Ancient Landforms

Ancient Landforms

Cliff Ollier

Belhaven Press
London and New York

© Cliff Ollier, 1991

First published in Great Britain in 1991 by
Belhaven Press (a division of Pinter Publishers),
25 Floral Street, London WC2E 9DS

British Library Cataloguing in Publication Data
A CIP catalogue record for this book is available from the British Library

ISBN 1 85293 074 8

For enquiries in North America please contact
PO Box 197, Irvington, NY 10533.

A CIP catalog record for this book is available from the Library of Congress

Typeset by Communitype Communications Limited
Printed and bound by Biddles Ltd of Guildford and Kings Lynn

Contents

Acknowledgements   vi

1   Landforms ancient and modern   1
2   Geological time and dating methods   5
3   Depositional landforms   17
4   Rivers and valleys   29
5   Weathering   48
6   Karst landscapes   68
7   Erosion surfaces   74
8   Plate tectonics and continental margins   104
9   Oceans and coasts   124
10 Volcanoes   136
11 Intrusive rocks   154
12 Faults and rifts   164
13 Mountains   180
14 Time, landforms and geomorphic theory   199
    References   214

Index   230

Acknowledgements

I thank the many people who have helped in the production of this book by discussion and argument, suggestion of items to include, provision of new material (sometimes unpublished), and sometimes by sheer encouragement. I especially thank my wife Janeta who helped not only with this book, but through the many earlier years of travel and research that lie behind its production.

I am grateful to the Department of Geography and Planning at the University of New England who provided the facilities for production of the book. In particular Rudi Boskovic, Graham Fry and Mike Roach drew the figures, and Megan Wheeler prepared the final text. Several people improved early drafts especially Bob Haworth and Russell Drysdale.

In previous books I have almost always managed to omit somebody who should have been thanked, or some reference that should have been acknowledged. In the present book I have tried to acknowledge all sources of information or help, but knowing my human weakness I apologise in advance to anyone who has been inadvertently left out. Experience has also taught me that books contain errors: for any here I accept sole responsibility. This is something of an iconoclastic book and in places is forcefully written, but I sincerely hope that nobody will feel personally attacked or offended.

Chapter 1
Landforms ancient and modern

This book is about the study of landforms, a science commonly called geomorphology or physiography, and it differs from almost all other books on the subject in the vast time-scale that is considered.

It is generally assumed that the landforms we see around us — the hills, valleys, mountains and plateaux — were largely formed in the Quaternary Period, that is in about the last two million years. The idea is furthered by the titles of some books and journals, such as *Landforms and the Quaternary*, a major abstract journal, or a leading Italian journal of earth science, *Geografia Fisica e Dinamica Quaternaria*. The Kirk Bryan Award is given annually by the Quaternary Geology and Geomorphology Division of the Geological Society of America. There is, of course, some truth in the assumption, and many small-scale features such as river terraces, sand dunes or coastal beaches really are Quaternary features, and often rather young Quaternary features at that. Indeed the trend in recent years has been to study actual processes operating on present day landforms, such as river channels or shingle beaches, and ignore the historical development of landscapes. But in reality, as I shall try to show, most of the major features of landscape, and some of the small ones, are older than two million years, and sometimes very much older.

How did this short-term view of landscape evolution come about? Part of the answer lies in the fact that at least in English-speaking countries, many geomorphologists are trained in geography rather than geology. They have little feel for geological time, they are not confident in geological techniques (especially reading geological maps), and they are often more interested in the relationship between landforms and the present environment than in the long-term evolution of landscapes. As Thomas and Summerfield wrote (1987):

A neglect of geology by geomorphologists has always been unwise.... Ostensibly this is because the 'past' is either irrelevant or unknowable, and inherited materials...can be accepted... without enquiry into their origins. But this view is limiting to the scope of research enquiry and for the inferences which can be made from field observation and measurement.

I should add that the neglect of geology is often due to ignorance, and many geomorphologists are unable to take full advantage of the information contained in geological maps.

Many geology courses provide very little training in geomorphology, and geologists are still often traditional hard-rock types who despise what they see as the trivia of surficial deposits and young landforms.

This book therefore has two purposes: to persuade geomorphologists that they need to enlarge their time-scale for a proper appreciation of landscape evolution; and to persuade geologists that the study of landforms is more than the last bit of carving on the geological column. Landforms are created on the same time scale as continental drift, sea-floor spreading, mountain formation and biological evolution, and the study of landforms puts serious constraints on geological and geophysical theories.

One reason why landscapes are commonly considered to be geologically young comes from the history of the science of geomorphology. Most of the early studies in geomorphology were in Europe and North America where it was realized, even before geomorphology had evolved into an independent study, that the landscape had been enormously modified by an ice age. Most of the landforms were therefore either glacial or post-glacial, which put them firmly into the Quaternary, the latest geological period which was at that time thought to be distinguished by its glaciation.

Thornbury (1969) wrote a book on geomorphology which at the time was outstanding, and enunciated several Fundamental Concepts of geomorphology. His Concept 7 was: 'Little of the earth's topography is older than Tertiary and most of it no older than Pleistocene.'

Yet even glaciated areas can have a long pre-glacial history as well (Figure 1.1). Furthermore, large areas of the world were never glaciated, and they do not have the fresh start to landscape evolution provided by glaciation in the northern part of the northern hemisphere. It was only by assumption that the Quaternary bias of northern studies was extended, for example, to Africa. River terraces were examined in Uganda, and related to hypothetical 'pluvials' or rainy periods, which were in turn related to the Quaternary glaciations of Europe, at that time thought to number four. The cultural cringe extended to Australia, where coastal ridges were at first thought to match a series of Quaternary sea-levels in the Mediterranean. When Europeans increased the number of proposed levels, new ridges were found in Australia to match. In time all these correlations proved false, as did the European theories on which they were based. Non-glaciated areas had no 'firm' dating techniques such as that allegedly (but often erroneously) provided by glaciation. The suspicion grew that such areas may be much older, but in the absence of good dating techniques it could not be proved.

Nevertheless, some imaginative thinkers such as Lester King related major landforms to global evolution, continental drift and mountain building on a time-scale that went back to the Tertiary and Mesozoic. His work was largely

Figure 1.1 Ruwenzori Mountains, Uganda. A view over Bujuku Lake, with Alexandra and Margherita peaks in the background, and below them the icefall at the head of the Stranley Plateau Glacier. The whole area was under ice in the last ice age. This photo was taken in 1956 and much of the glacier shown has now disappeared. But to explain the geomorphology of the area fully it is necessary to include uplift of the Ruwenzori block which probably occurred in the Miocene, perhaps 10 million years ago (Photo: J.F. Harrop)

neglected, perhaps because it appeared in the 1960s, just when a 'New geomorphology' was being espoused which was allegedly independent of time!

Major advances in putting ages to landscapes came when potassium argon dating enabled basalts to be dated. In south-east Australia there are many landscapes with Tertiary basalt, and when good dates were available it became possible to put landscape evolution into a dated sequence, and work out secondary features such as erosion rates, the age of weathering profiles, and many other aspects of landscape evolution. Australia is lucky in having these basalts which were erupted throughout the Tertiary over a wide area. Other places where volcanism helps to decipher landscape history are East Africa and the western United States. Some places where it would be highly desirable to date the landscapes, such as Africa and South America, do not have such convenient basaltic time markers. Perhaps for this reason Australian geomorphologists had a flying start in the study of ancient landforms, and many examples in this book come from Australia. The overemphasis on Australia is not just because I know it better, but because more has been written about Australian ancient landscapes than those of any other continent. In this book I have tried to use examples from other continents, but an Australian bias clearly prevails. I hope that in future many more examples will be available from other continents.

This will certainly happen, and is already under way. As we shall see, even in thoroughly glaciated Sweden there are landforms that can be traced back to the Cretaceous, perhaps a hundred million years ago. Examples will multiply, and perhaps the time will arrive when the great age of landforms will be accepted as readily as the youth of landforms is accepted at present.

In this book I hope to demonstrate the 10th Fundamental Concept of Geomorphology enunciated by Thornbury (1969): 'Geomorphology, although concerned primarily with present-day landscapes, attains its maximum usefulness by historical extension.'

Chapter 2
Geological time and dating methods

There are many ways of dating events, some obvious and some obscure. The layman is often suspicious of geological ages, wondering, 'How can they know?' and often finding the creationist or pseudo-science explanations of world history as satisfactory as scientific explanations. The following chapter attempts to explain how dates are obtained. There is a perspective of time, and events closer to us can be dated more accurately than those further back in time, but then we usually need less accuracy when dealing with remote events.

Observation and historical dating

The easiest way to date a geomorphic event is to observe it happen. The flank collapse of Mount St Helens, for example, occurred on 18 May 1980 and for many purposes could be dated even to the minute or hour. The catastrophic landslide of Mayunmarca in Peru, 8 km long, took place on 25 April 1974. Such observations are limited to the lifetime of the observer. A longer time scale can be gained from historical observations. The eruption of Vesuvius in AD 79 is well-documented. The change in course of the Huang Ho (Yellow River) is well known, as it killed a quarter of a million people in 1851. More vague, but nevertheless well-documented is a little ice age that occurred in Europe between the Middle Ages and the twentieth century (Grove, 1988).

Earlier records are less definite. Describing the Acropolis in Athens Kritias recorded, 'To begin with, the Acropolis was not then as it is now. At present it has been washed bare of soil by one night of heavy floods', which possibly happened in 275 bc (Kukal, 1984, p.20). Floods of the Nile were recorded from the time of the Pharaohs. Such records indicate the changing circumstances at the earth's surface for several thousand years at most, and fall far short of the time-scale of this book. What techniques can extend our knowledge further back?

Relative dating

Superposition

In many locations successive layers of sediment are deposited, as for instance on the sea-floor, in lakes, or in caves. In many places the process is still active and a new layer is deposited on top of earlier ones. At some time in the future it may be overlain by a still younger layer. The youngest layer is always on top, and any layer is younger than the one beneath. The same rule applies if the strata are raised above sea-level, or if a complicated sequence of uplift, erosion, folding and subsidence is present (Figure 2.1). This simple relationship has been elevated to the status of a law — the law of superposition. It is basic to the study of layers of sediment (strata) and the science of stratigraphy. A modern textbook dealing with the principles of stratigraphy is Schoch (1989), but elementary introductions to the subject are given in most textbooks of geology.

In some places there are only a few layers, in others there may be thousands. In Ecuador a borehole through a peat deposit revealed a succession of layers in a swamp which were analysed and dated, revealing 27 major coolings (ice ages) in the past 3.5 million years (Hooghiemstra, 1984). Williams (1989) described a bore with thousands of layers, recording tidal and other cycles of deposition in the Precambrian (about 700 million years ago), all based on the law of superposition.

At any given place in the world, the number of strata preserved and observable must be limited, but it is possible to move from place to place and correlate distinctive strata. Strata may be recognized by the assemblage of fossils, and it was this discovery which really marked the beginning of historical geology — the unravelling of the history of the earth by the study of rocks and fossils which could all be put in the right relative position in time. Other features such as chemical, magnetic or mineralogical features can also be used to correlate strata. In Quaternary studies the reversal of the earth's magnetic field at about 700 000 years ago is often an important datum. Mineral correlations are used in the science of tephrochronology, the study of volcanic ash. It is found that many volcanic ashes from individual volcanic eruptions are far-travelled, and can be traced and correlated over wide areas. In places like New Zealand where eruptions have been frequent, the relative ages of hundreds of ash showers have been deciphered.

By analysing local and regional correlations over many years, geologists have been able to work out the main features of the stratigraphic history of the earth, and for ease of communication have divided the 'geological column',

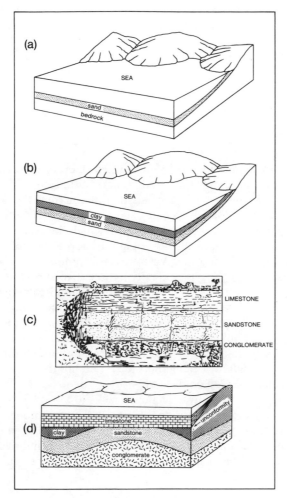

Figure 2.1 Some principles of stratigraphy.
(a) Deposition of sand on the sea floor. The sand, derived from the land by erosion and transport, is younger than the bedrock beneath.
(b) In a later stage sand deposition is replaced by clay deposition. Since it is above the sand, the clay is younger.
(c) Strata exposed in a quarry. The original sediments were pebbles (turned into conglomerate), sand (sandstone) and shelly deposits (limestone). Using the law of superposition we can see that the conglomerate is the oldest rock represented, and the limestone is the youngest. The area has been uplifted above sea level since the strata were deposited, but we cannot see from the picture whether there has been any natural erosion.
(d) Although not geologically very complicated this figure illustrates a minimum geological history as follows: deposition of pebbles, sand and clay, in that order, which became conglomerate, sandstone and clay. These strata were folded, uplifted and eroded to a plain. Since then there has been subsidence, and deposition of limestone above an unconformity that separates the unfolded limestone from the folded strata beneath.
(Source: C.D. Ollier, *Tectonics and landforms*, Longman, 1981)

which is the total list of possible strata, into several divisions. These are shown in Figure 2.2. In any particular part of the earth certain sections of the table will be more relevant than others, and finer divisions will be needed. There are therefore many local names in use, but all can be correlated with the main geological column.

Cross-cutting (intrusions)

Granites and other 'plutonic' rocks are formed deep in the earth, and have rock-relationships different from sedimentary strata. Deep in the earth metamorphic rocks may slowly recrystallize under the effects of heat and chemical migration, and slowly change to granite (though sometimes retaining a 'ghost stratigraphy' of the earlier rocks from which the granite was created). The granite may move upwards, essentially by replacing pre-existing rocks rather than pushing them aside. Vast areas of granite called batholiths are formed in this way. Granite is generally lighter than surrounding rocks and tends to rise, and some parts of a batholith may rise like chimneys, pushing aside the overlying rocks, or 'raising the roof' by pushing it up. Such granites are called plutons. For the time element, the most important point is that granites have to be younger than the rocks they intrude.

Emplacement of granite usually bakes (metamorphoses) the surrounding rocks, often to a tough rock called hornfels. The hornfels has to be younger than its unmetamorphosed equivalent.

Some igneous rocks are intruded along cracks to form dykes. A dyke must be younger than the rock it intrudes. If one dyke cuts through another, the unbroken dyke must be younger than the one which has been cut. All these relationship must be summarized as the law of cross-cutting relationships. In igneous rocks, the one that cuts across rocks is younger than the one it interrupts.

Fault sequences

Faults are cracks across rocks along which there has been relative movement. In the simplest case a nearly vertical fault breaks the rock, and one side is downfaulted relative to the other. The fault has to be younger than the rocks that are faulted. In time younger sediments may come to lie over the fault. If they are not affected by the fault they are clearly younger than the fault, and indeed put a limiting age on the date of faulting.

Faults may also move sideways (strike-slip faults), possibly breaking fences, roads, railways or valleys. A valley that has been moved sideways is older than the fault. A valley that runs straight across a strike-slip fault has a course that is younger than the fault. In some instances faults move over a long

ERA	PERIOD	EPOCH	M.Yrs Ago
CENOZOIC	QUATERNARY	RECENT- about 10,000 years	
		PLEISTOCENE	1.8
	TERTIARY	PLIOCENE	5.5
		MIOCENE	22.5
		OLIGOCENE	38
		EOCENE	54
		PALEOCENE	65
MESOZOIC	CRETACIOUS		135
	JURASSIC		190
	TRIASSIC		225
PALEOZOIC	PERMIAN		280
	CARBONIFEROUS		345
	DEVONIAN		395
	SILURIAN		440
	ORDOVICIAN		500
	CAMBRIAN		570
	PRECAMBRIAN		

Figure 2.2 The geological time scale. The Cenozoic is sometimes called the Cainozoic. The units are sometimes split up into Lower, Middle and Upper, or lower, middle and upper. Capital letters are used for formal stratigraphic usage, lower case letters for informal usage. The same applies to early, middle and late. (Source: C.D. Ollier, *Tectonics and Landforms*, Longman, 1981)

period and landforms are successively transported sideways. Figure 2.3 shows a series of river terraces in New Zealand affected by faulting. The oldest terrace, affected by all subsequent movement, has moved the most, but the youngest terrace is affected by only the last movement.

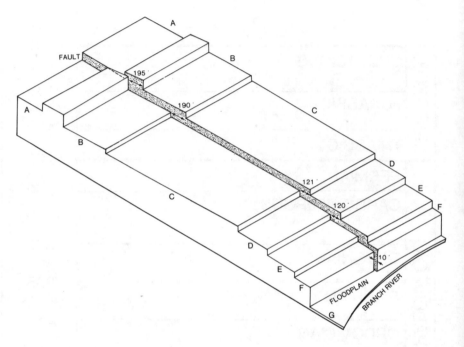

Figure 2.3 Diagram showing the progressive offset of terraces and channels of the Branch River, New Zealand by the Wairau Fault. The dimensions of the offsets are in feet (after Lensen, 1968).

Erosion sequences

Erosion carves valleys and hills in pre-existing rocks, and so must be younger than the rocks. A valley on a 50-year-old volcano cannot be more than 50 years old; a valley cut through Cretaceous limestone cannot be older than Cretaceous. It can of course be very much younger.

In time erosion can wear away most of the hills and produce plains. The age of a plain is younger than that of the rocks it is cut across. The plain shown on Figure 2.3 is cut across Cretaceous sediments, and so is not older than Cretaceous and is probably post-Cretaceous. Vertical erosion by rivers may cut into a planation surface, forming a series of valleys, and these in turn may widen to produce another, lower, planation surface. The lower erosion surface is younger than the higher one. It is generally assumed that there is no way that the high erosion surface can be produced after the formation of the lower one. (An exceptional interpretation of erosion surfaces by Battiau-Queney is described on p. 200).

Figure 2.4 An erosional plain cut across Cretaceous sediments, and so no older than Cretaceous. Coober Pedy, South Australia. (Photo: C.D. Ollier).

Marine terraces

Coastal erosion and deposition leads to a different set of landforms from that of ordinary river activity. If sea-level remains stable for some time a terrace may be formed at around sea-level: erosion works down to sea-level, and deposition (of deltas, coral reefs, etc.) builds up to sea-level.

If there is a relative fall in sea-level, the terrace will be left high and dry, and a new terrace will start to form at the new sea-level. The process may be repeated many times, giving rise to a flight of terraces. In such flights of terraces, the higher ones will be older than the lower ones.

Retreat sequences

A glacier transports a load of debris, some of which is dumped at the front of the glacier as a terminal moraine. If the glacier retreats, the moraine is left in its original position, and a new moraine is deposited at the new glacier snout. If the glacier should retreat to a third position, a third moraine will be formed. The oldest moraine will be the furthest downstream.

A similar situation can occur on coasts. A beach ridge may be formed at about sea-level. With relative fall in sea-level the ridge may be abandoned, and a new beach ridge formed at the lower sea-level. In time a whole sequence of beach ridges may be formed. The oldest ridge will be the furthest inland; the youngest will be at the present coast.

Drainage modification

Rivers and their tributaries make distinct patterns as seen in plan (described in Chapter 4). Initially the pattern is usually simple, but there may be complications by such modifications as river capture or diversion. Any river modification must be younger than the original simple course. Thus the original course of the Barron River (Figure 4.9) must be older than the river capture which diverted it down the Barron Gorge. The course of the Simeto River around Mount Etna (Figure 10.8) must be younger than the ancient course which was obliterated by the eruption of Mount Etna.

River terraces

A river floodplain may be essentially built by deposition of alluvium, or cut across rock by lateral erosion of the river. In either case, if the river then cuts down to a deeper level, an incised valley is formed and remnants of the old floodplain are left as river terraces. In some places repeated periodic incision has created a whole series of river terraces, and in such cases the highest terraces are the oldest.

Weathering sequences

The course of weathering is more complex than some of the items considered so far, but in general, weathering increases with the passage of time. In Western Samoa, for instance, the oldest volcanic rocks are weathered to a stone-free condition; those of middle age are weathered to a few centimetres; and the youngest volcanic rocks are still hard and fresh. Similarly in glacial moraines, the oldest have the most weathering, the youngest the least. Beach ridges, river terraces and similar landforms show the same tendency.

Soil sequences

Soil is formed at the earth's surface by a range of processes including organic accumulation, mixing and sorting; leaching and precipitation; eluviation and illuviation; cheluviation; and perhaps other processes. The result is a soil profile consisting of several layers or horizons.

Several factors influence soil formation, including parent material, climate, topography, organisms and time. The time factor is revealed very clearly when a group of landforms of the same origin but of different age are present in the same area. A sequence of river terraces will have been influenced by soil-forming processes for different lengths of time. So will other successive features such as beach ridges, moraines or dunes. The older the landform, the greater will be the soil development, which may be expressed in soil depth, more intense mineral alteration, accumulation of organic matter, or other features. Some examples are given in Chapter 5. Most soil sequences are too young to be useful in the study of long-term landscape evolution.

Absolute dating

A number of techniques are available which can potentially provide absolute dates. Only a few will be mentioned here.

Carbon dating

This method can be applied to material containing carbon, such as charcoal, bones, shells, or wood. It is reliable to about 35 000 years, but becomes increasingly suspect with greater age. Such dates are too young to be of interest in this book.

Potassium/Argon dating (K/Ar dating)

Potassium occurs in reasonably large amounts in all common rocks. One of the isotopes of potassium — potassium-40 — is radioactive and may decay to produce argon-40. When igneous rocks form, all gases present are generally released, so that the rock starts with potassium-40 but no argon-40. Argon-40 will therefore accumulate in the rock, depending on the original potassium content and the time that has elapsed, thus providing a dating method. The main problem with the method is loss of argon. This method is especially valuable for dating volcanic rocks and can be applied to specimens from a few tens of thousand years old to very great geological ages.

Palaeomagnetism

Both sedimentary and igneous rocks may acquire a magnetic signature at the time of formation, and retain it later. Magnetic dip can be determined from old rocks to provide the latitude at which a rock was formed, and if enough background work has been done, this may be used to determine the age of the rock.

Another feature of earth magnetism is that sometimes the magnetic field is like that of today (called 'normal') but at other times is completely reversed. Determination of magnetic direction can sometimes be used to provide ages. The last major reversal was 700 000 years ago. A series of beach ridges in South Australia have normal magnetization, from the present day active beach to some tens of kilometres inland. The dunes further inland have reversed magnetism and so are older than 700 000 years (Cook et al., 1977).

Fission track dating

Fission tracks are tracks seen under the microscope in minerals such as apatite and zircon that contain a trace of uranium. They are produced as a result of radiation, depending on the amount of radioactive elements in the specimen (which can be measured), and the tracks are annealed by heat. The method can be used to date the last time the rock was hot enough to destroy pre-exising fission tracks, which is simple in an igneous intrusion which has simply cooled down, but can be difficult if complex scenarios such as the passage of a thermal event are involved.

Fission track dating is sometimes equivocal, and other data are needed to arrive at an interpretation. In South Africa for example, fission track dates fall into two groups — an old group of over 100 million years, and a younger group with ages close to 70 million years (Gleadow and Duddy, 1987). One explanation involves removal of over 3 km of overlying rock from both highland and lowland (very improbable), and the second explanation requires an early Cretaceous thermal overprint under a much shallower cover of overlying rocks (more probable). Similarly in south-east Australia, fission track dates are markedly younger near the coast, which can be interpreted as a thermal effect at the time of formation of the continental edge, or erosion of several kilometres of rock (Dumitru et al., in press).

Physical methods (e.g. fission track data, vitriance reflectance) that purport to show large amounts of erosion are often at odds with geomorphic evidence. Friedman (1987) reports that many techniques applied to rocks of the northern Appalachians suggest removal of 4.3 to 7 km of rock from the present ground surface. All geomorphic techniques suggest very much less. Similarly, numerous authors have postulated thicknesses of several kilometres over the presently exposed rocks of the Sydney Basin, Australia, but Branagan (1983) believes the stratigraphic and geomorphic evidence will allow considerably less than one kilometre. Perhaps the estimates of palaeotemperature gradients are wrong.

Many other absolute dating methods are available, including thermal luminescence, uranium-lead dating, uranium series dating, and more.

Indirect dating methods

Several indirect methods of dating are available, but all have some disadvantages and depend more than we should like on assumptions. The sea-level has changed through time, and its course is to some extent charted. High sea-levels occur in interglacial periods, low sea-levels in glacial periods. However, the exact number and age of former sea-levels is hard to determine, and is not constant around the world. Tectonic activity has created too many complications. Nevertheless, with some reservations, sea-level provides some useful dates, especially for the last interglacial of about 125 000 years.

Sea-level has also been determined through geological time by complex geological methods, and a series of curves commonly known as Vail curves have been determined to show the change in sea-level for the past 230 million years (Haq et al., 1987), and indeed for the whole Phanerozoic.

Climate has also changed through time, and may give clues to age. For over fifty years geomorphology was dominated by the idea that there were four main ice ages, with an especially intense 'Great Interglacial' in the middle. This is now known to be quite untrue, but serves as a warning against a too-simple view of past climates. On a longer time-scale climate has changed severely, with ice ages in the Late Precambrian, Permian and Quaternary; an arid period in the Triassic; and a period of warm wet climate in the Cretaceous.

Many techniques are now available for determining the rate of erosion. They include the measure of bedload, suspended load and solute load of river; measurements of hillside creep; measurement of coastal retreat, and many more. It is also possible to measure rates of tectonic movement by repeated survey, by disturbance of man-made features such as canals, by examining raised beaches, and many other methods.

After thousands of such measurements the following averages seem to be somewhere near the mark:

Rate of erosion of lowlands	50 B
Rate of erosion of highlands	500 B
Rate of uplift	1 000 B
Rate of sideways movement	10 000 B

(B symbolizes Bubnoff unit, and is 1 mm per 1000 years or 1 m per million years)

However, averages are just that, and do not reveal the extremes very well. In Western Australia rates of erosion of less than 1 B have been suggested, and some faults have moved over a metre in a few seconds, giving enormous rates if extended (unrealistically) over a long time. Nevertheless, a knowledge of average rates does give a standard against which other observed rates can be compared, and in some circumstances rates can be translated into time to put age constraints on landforms.

Problems of dating

Some actions at the earth's surface are virtually instantaneous, such as a volcanic eruption or a meteorite impact. There is no theoretical problem in providing a date, even if it may be difficult in practice.

Other features are not so crisp. When does a soil start to form? When, if ever, does a soil stop forming? When does a climate start to change? When does a slope start to retreat? These are all examples of rather indefinite events. Although in principle they can be dated, there is no point in trying to be too precise.

Another problem arises when dates are correlated. With the onset of aridity a river may dry up first at its mouth and later upstream, so a series of lakes along the river may dry up in turn. In this case it would be pointless to correlate the drying event from lake to lake — the time was different for each lake.

As a large escarpment retreats by erosion, a plain at its foot is enlarged. The part of the plain closest to the escarpment has been formed only recently, but parts of the plain remote from the escarpment may be millions of years old. And yet the plain is continuous. It is of different age in different places and is said to be diachronic (across time).

This brief account of methods of determining age is rather arid in isolation, but some of the concepts are necessary to understand the data presented in later chapters.

Chapter 3
Depositional landforms

Depositional landforms are those made by deposition of sediment, such as beaches, alluvial plains or sand dunes. Sediments may be dated by contained fossils, which is the basis of the science of stratigraphy. But sediments are not only deposited in layers or strata: they sometimes make actual landforms which may be dated by contained fossils, or other means.

Sand dunes

Because sand is so easily mobilized, dunes of great age are rare. Nevertheless, sand dunes are preserved in many parts of the world, even when the processes that gave rise to them are no longer active. There are two main kinds of dunes: desert dunes formed in arid areas where wind is dominant and a sand supply is available, and coastal dunes formed where coastal processes combine to produce a sand supply in an area of onshore winds. Many dunes relate to the last ice age, when aridity increased in the world and many dunes that are now fixed were actively growing and migrating, and when the sea-level was low. In Australia no pre-Quaternary dune fields are known, and most are Upper Quaternary. Sea-level was low during the ice age, exposing broad flats along the coast in suitable areas that were covered in sand which could be built into coastal dunes.

Parts of the Queensland coast, in north-east Australia, had an ideal combination of large sand supply, broad coastal flats, and strong onshore winds, and here the world's highest dunes were formed (Coaldrake, 1962). The dunes extend from 70 m below sea-level to a height of almost 300 m. It is thought the dunes have been building throughout the Quaternary and probably longer.

High dunes are also associated with ancient beach ridges, including the Ooldea Range, a sand dune 650 km long and 40 to 180 m high, which lies 250 to 300 km inland from the present coastline. The Range is of Eocene age, perhaps 34–37 million years old (Benbow, 1990).

Loess

Another aeolian deposit is the dominantly silt-sized loess, which attains thicknesses of thousands of metres in China and is the basis of a whole landscape over a very wide area. The loess goes back to at least 2.4 million years (Liu and Yuan, 1987).

Beach ridges

In South Australia part of the present coast consists of a beach ridge over 125 km long, behind which is a lagoon called the Coorong. Inland from the lagoon is another beach ridge, clearly older than the active one, and further inland is another, and then another.... These ridges can be traced inland for hundreds of kilometres into the State of Victoria (Figure 3.1). The seaward ones have been shown to be Quaternary and the inland ones are Tertiary (Cook et al., 1977), the dates being based on fossil content and palaeomagnetism. The oldest ones of this sequence are Miocene.

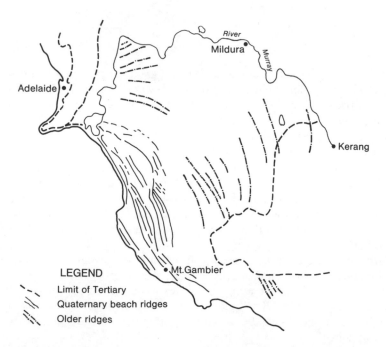

LEGEND
Limit of Tertiary
Quaternary beach ridges
Older ridges

Figure 3.1 Beach ridges of south-east Australia. Solid lines Quaternary, dashed lines Tertiary. (C.D. Ollier)

Similar beach ridges are also present north of the Nullarbor Plain, extending from South Australia into Western Australia, and are of Eocene age, about 37 million years old (Benbow, 1990). At present the aridity of the region helps to preserve the features, but it has not been arid for the entire 37 million years of their existence. Underground drainage and the loss of surface streams is probably equally important for the preservation of such ancient beach ridges.

River sediments

'The literature of ancient river systems, especially those predating the late Tertiary, is very widely scattered and not voluminous. All too commonly, it is the indirect product of other study' (Potter, 1978). Potter's article is a milestone in long time-scale geomorphology, even though he is mainly a sedimentologist.

River sediments such as floodplains and terraces may be dated by various means, but this is generally only useful in the Quaternary Period. Some major rivers have flowed in their courses since the Lower Tertiary (e.g. the Kagera, Uganda, see p. 40), but do not contain datable sediments. Ancient river systems can be traced even in glaciated areas, and McMillian (1973) reconstructed the palaeodrainage of much of Canada with drainage towards a Tertiary basin between Canada and Greenland, so the drainage must also be Tertiary.

McCartan et al. (1990) described upland gravels and sand sheets near Brandywine, Maryland, which are an emerged part of the Atlantic coastal plain and bevel Cretaceous and Miocene sediments. Within the surficial materials they found a clay lens which was a deposit in a cut-off distributary on a braided stream system. Abundant fossils showed it had a late Miocene age.

In the mountains of southern and central Wyoming, local relief at the end of the Eocene was as great as it is today, or greater, as indicated by large palaeovalleys that were cut in Eocene time and later filled with Oligocene White River Formation (Evanoff, 1990).

An exceptional example comes from the Norseman area of Western Australia. The area is arid at present, but chains of salt lakes and saline soils mark the course of ancient valleys, and cross-sections are exposed in large open-cut mines, as at Central Norseman (Figure 3.2). The lowest sediment in the valley is terrestrial, and contains leaves, fruits and pollen indicating an Eocene age of about 40 million years. These are overlain by marine sediments containing sponge spicules and sharks' teeth, also about 40 million years old. Evidently a marine incursion caused by a rise in sea-level extended from the south for hundreds of kilometres up the valleys, apparently without overflowing on to the surrounding plain. We not only know the age of the river sediments, but can conclude that by 40 million years ago the valleys were not only in existence, but they were incised in a very flat plain and had gentle gradients. Obviously the age of the plain is much greater than Eocene (Ollier, Chan et al., 1988).

Saucy Creek, once one of the headwaters of a palaeo-Murrumbidgee in

Figure 3.2 A valley exposed in a mine at Central Norseman, Western Australia. The lower part of the valley fill is terrestrial sediment with fossil plants: the upper fill was deposited by a marine transgression, and contains sponge spicules and shark teeth. (Photo: C.D. Ollier).

southern New South Wales, was blocked by the Monaro Volcano in the Paleocene and became a lake in which Paleocene sediments accumulated (Ollier and Taylor, 1988).

Near Bendigo, Victoria, local plateaux are covered in coarse debris which is interpreted as being the sediment laid down by huge rivers that meandered across a gentle plain (Figure 3.3). The sediments are older than the early Tertiary lava flows which filled later valleys, and they are thought to be Cretaceous (Williams, 1983). In Western Australia the gravels of the Fortesque River appear to be of great age. Although not fossiliferous, correlation with neighbouring sediments suggest a Cretaceous age.

Tin placer deposits in northern Tasmania were formed by episodic recycling dating back at least to the Permian (Yim, in press). Heavy mineral species demonstrably derived from granitic and basaltic source rocks were used to identify placer formation events. Deep leads were created when lava flows covered the alluvium, first in the Middle Eocene (about 47 million years ago) and again in the Middle Miocene (about 16 million years ago).

Care is needed to distinguish simple old landforms from those which are exhumed after burial under younger strata, and so do not have a continuous history as landforms. On the Mendip Hills in England, for example, the Carboniferous limestone was eroded by valleys which were subsequently filled by Triassic sediments. These are now exposed again, but the whole area was possibly covered by Mesozoic strata and re-exhumed at a later date, so the age of the valleys today is not indicated by the age of the sediments they contain. Similarly, some of the valleys of the Kimberley Plateau in Western

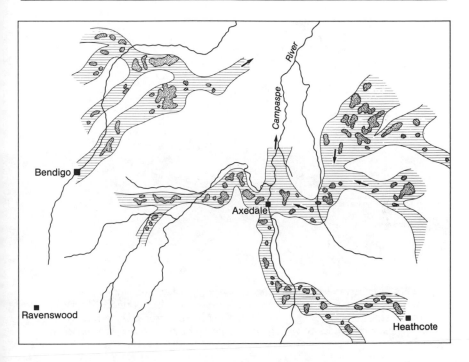

Figure 3.3 Remnants of probably Cretaceous gravel on plateaus (stipple) in the Bendigo area, Victoria, Australia, and reconstructed drainage pattern by G.W. Williams, using also the elevation of the gravels, not shown here. The gravels consists largely of quartz pebbles, but those in the east also contain chert from Cambrian outcrops in their catchment.

Australia are exhumed from beneath a cover of Precambrian glacial sediments (see p. 24).

Martin (1975) described palaeodrainage in south-west Africa, where valleys up to 1000 m deep were filled with Permo-Carboniferous glacial deposits and later by Cretaceous basalts. The palaeodrainage system is oriented towards the present African coast, suggesting the presence of a proto-South Atlantic Ocean in the Late Palaeozoic.

Perhaps the oldest river sediments still preserved at the earth's surface are in the Davenport Ranges, northern Australia. Here an ancient erosion surface is preserved on steeply dipping quartzite ridges, and valleys are cut into the softer rock between. Alluvial terraces are preserved in these valleys, which were originally mapped as Tertiary, but it later transpired that these gravels interfingered with fossiliferous marine sediments further south (Figure 3.4). The fossils are Cambrian, so presumably the terraces are also of Cambrian age (Stewart et al., 1986).

Swann (1963) reconstructed an ancestral Mississippi and showed that a river has been oriented across the North American continent since the Early

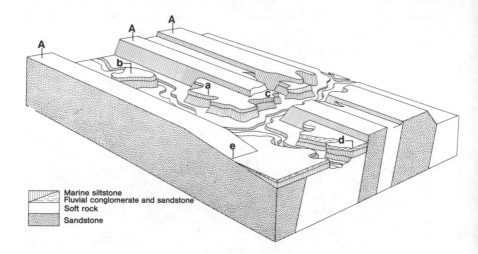

Figure 3.4 River terraces of the Davenport Ranges, Northern Territory, Australia. Remnants of the Ashburton Surface (A) are preserved on ridges of steeply dipping Proterozoic sandstone. Flat-lying Cambrian sediments are preserved in terraces (localities a, c and d) and a mesa (locality b). Cambrian strata lap onto the Ashburton Surface at locality e.

Carboniferous, following an axis which had its origins in the Late Precambrian, but this does not mean that the present Mississippi is an ancient river. Much of it is very young indeed. Nevertheless, to understand the modern river it is necessary to know that it had precursors in the same area over a very long period (Figure 3.5). Part of the evidence for the persistence of this river system is found in the palaeochannel system shown by the region's pre-Mississippian unconformity. More fundamentally, the river follows a

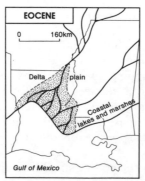
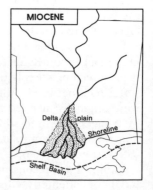

Figure 3.5 Evolution of the Mississippi River since the Jurassic. (Simplified from Potter, 1978).

reactivated rift that had its origins in the Precambrian (Potter, 1978). The Michigan–Mississippi river system has a longevity of about 250 billion years!

Deltas

Deltas are bodies of sediment built offshore from river-mouths. Many are still actively growing, but what concerns us here is the age of deltas. Any delta must be younger than the sea-floor upon which it lies, so since the Atlantic Ocean did not appear until the Jurassic, there should be no pre-Jurassic deltas on its edges. Deltas of the Gulf of Mexico coast of the United States date back to the early Tertiary (Fisher and McGowen, 1967).

The Niger delta is one of the largest deltas on the Atlantic margin, and has been growing since the Lower Tertiary at least (Evamy et al., 1979), and there was a proto-Niger in the Cretaceous (Frankl and Cordry, 1967). The significant feature is that the Niger has been delivering sediment to this delta at least since this time. The Niger delta is sufficiently large to require 'removal' when pre-drift continental fits are reconstructed.

Another huge delta is that of the Ganges–Brahmaputra (Moore et al., 1974; Wezel, 1988), a submarine fan over 2000 km long. This covers an enormous area, and the distal parts are folded (Figure 3.6) — possibly by the most distant effects of the Himalayan orogeny (Wezel, 1988), but perhaps by gravity sliding. The oldest known sediments in the delta are thought to be of Upper Cretaceous to Paleocene age.

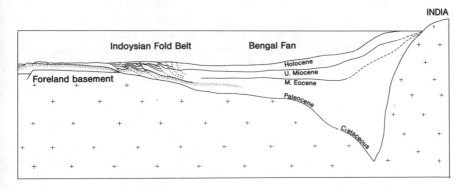

Figure 3.6 Cross section of Bengal Fan and Indoysian fold belt (Simplified from Moore et al. 1974).

In relation to global tectonics and drainage it is interesting to note that twenty-five of the world's largest deltas are found on passive margins (Potter, 1978), suggesting that deltas and global tectonics share the same time-scale.

Glacial sediments

Glaciers and ice sheets deposit characteristic sediments including till, an unsorted and unstratified sediment with a wide range of grain size from boulders to clay. This often has characteristic landforms, such as the ridge known as a terminal moraine that marks the farthest extent of a glacier. Moraines have been used to date glacial events and are especially useful for dating within the Quaternary. Even then care is required. An early idea based on moraines in an Alpine valley was that the Quaternary had four ice ages, a notion that dominated Quaternary science for at least fifty years and is still prevalent, even though it is now known that there are as many as twenty-seven major coolings in the past 3.5 million years (Hooghiemstra, 1984; Flenley, 1984).

Older till (known as tillite when hardened into rock) is well known from ancient times, especially the Carboniferous-Permian glaciation. Old tillites are frequently associated with actual landforms, especially glacial pavements that bear the scratches, gouges and polish of ancient glaciers. These are generally exhumed, but in a few instances relate closely to present valleys.

In northern Victoria the Loddon Valley follows a course which runs along an ancient valley that was once a Permian glacier. The old till was eroded, and Tertiary sediment was deposited in the same valley, and this was capped by a Tertiary basalt. The modern valley follows the same course, but of course is not directly a Permian valley (Macumber, 1977).

Similarly, also in northern Victoria, the Ovens valley follows a tectonic depression which was glaciated in the Permian. The till was largely eroded and replaced by Tertiary gravel, which was in turn eroded and covered in places by Quaternary sediments to form the landscape of today (Craig, 1984). Although it would be wrong to think today's landscape is the same as that in Permian times, it is certainly true that the broad features of the landscape can be traced directly to that time. Such landscapes which keep reappearing can be termed resurgent landscapes. Macumber (1977) pointed out that many north-draining rivers in northern Victoria, tributaries of the Murray, flow along fault angle depressions formed in the Permian and filled by Permian sediments.

On the Kimberley Plateau of Western Australia an erosion surface was striated by the Sturtian glaciation of about 700 million years ago, and covered by glacial till. The present striated surface is exhumed (Figure 3.7), but there was probably never any cover other than the Precambrian glacial and associated sediments (a conclusion based on regional palaeogeography). The significant thing is that although the individual glacial pavements are clearly exhumed, the evidence shows that the Kimberley Plateau was already in existence in Precambrian times (Ollier, Gaunt and Jurkowski, 1988).

Pseudo-glacial deposits are sometimes recorded. The Yakataga Formation of the Gulf of Alaska contains a 5 km-thick record of marine sediment from which a detailed picture of climate, landscape, sea-level, tectonic and oceanographic changes is being established for the past 6 million years. The White River Formation, a terrestrial correlative, consists of diamictite (unsorted sediment, once described as tillite) and conglomerate, interbedded

Figure 3.7 Glacial striations of Precambrian age Kimberley Ranges, Western Australia. (Photo: C.D. Ollier).

with lavas. It consists of debris flows and alluvial fans with a history spanning millions of years (Eyles and Eyles, 1989).

Cave fill

Cave sediments must be younger than the cave in which they are found. Many cave deposits are known, and most turn out to be of Quaternary age, but occasionally older sediments are found.

The Timor Caves in New South Wales contain Tertiary basalt, which according to some investigators flowed into pre-existing caves. The Bungonia Caves of New South Wales were developed and filled with sediment in the Early Eocene, the present-day caves being largely formed by exhumation (Osborne and Branagan, 1988).

In England some caves of the Mendip Hills have fill of Triassic sand, sometimes very fossiliferous, so at least some caves had formed in the Carboniferous Limestone by Triassic times. Further examples are given in Chapter 6.

Marine sediments

Marine sediments are usually easy to date by their fossil content, and if they make a distinctive depositional landform it can be dated accordingly. Without some change in sea-level, the geomorphic feature will still be a marine sediment, like the Great Barrier Reef of Queensland which is apparently entirely of Quaternary age.

With uplift, a marine terrace can be made. The Nullarbor Plain, which is about 200 m above sea-level and very flat, is essentially an uplifted Miocene sea-floor, although up to 65 m has been eroded from the southern edge (Lowry, 1970). The surface is varied by a few sink holes and possible beach ridges, and a certain amount of weathering and stripping (Lowry and Jennings, 1974; Twidale, 1990).

Elsewhere flights of marine terraces, mostly coral, are created, and in general the higher the terrace, the older it is. In the Finchhafen Peninsula of Papua New Guinea these have been traced to heights of 2 500 m (Chappell, 1974). The higher ones have not been dated, but are thought to be pre-Quaternary.

Ancient coasts and marine sediments may be preserved well inland after uplift. At Grange Burn, Victoria, an old coast with a shore platform and rock stacks and fossiliferous marine strata about 5 million years old is exposed about 60 km from the present coast.

Marine sediments can occasionally indicate something about pre-existing landscape features. In southern Sweden, for instance, Cretaceous sediments with fossil oysters mark a Cretaceous shoreline (Lidmar-Bergstrom, 1988). This would not be so remarkable but for the fact that some of the oysters are growing on exposed corestones in a deeply weathered granite, showing that deep weathering had already occurred in the area by Cretaceous times.

Depositional basins

At the small scale a depositional basin may be merely a lake. For instance the Lough Neagh Basin in Northern Ireland. This contains sediment up to 350 m thick, where it overlies 7 m of weathered basalt. The Lough Neagh Clays are of Oligocene age (Smith and McAlister, 1987).

At the largest scale are the major depositional basins. At present these are the oceans (especially the continental margins) and internal drainage areas such as the Mediterranean Sea and the rift valleys. Modern sedimentary basins are more the concern of marine geologists than geomorphologists, but it is occasionally possible to integrate the two studies. Ancient sedimentary basins are also of interest to geomorphologists so long as they work on the appropriate time-scale. Changes in drainage pattern in the catchment will be reflected in the nature of the sediment deposited in the basins.

In south-east Australia and surrounding regions, before sea-floor spreading and continental drift had isolated the present landmass of Australia, a broad area of Pacifica (a postulated super-continent that extended Australia to the east) drained towards the Great Artesian Basin (strictly, the Eromanga–Surat Basin). Rivers brought sediment from the east, and their modern equivalents are preserved as the Clarence, Hunter, Shoalhaven and others (now reversed and draining to the east coast). Other rivers brought sediment from the south, and their traces are still preserved as the Murrumbidgee, Ovens, Loddon and others. Rifting followed by sea-floor spreading beheaded most of the rivers that came from the east, reducing sediment supply to the Great Artesian Basin from this source. This happened at about 80 million years ago, when a new coastal margin to Australia was created (Ollier, 1992).

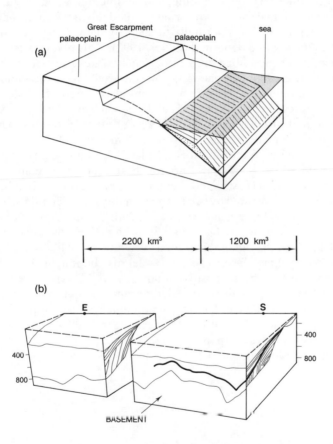

Figure 3.8 (a) Diagram of the coast of New South Wales. The palaeoplain downwarped to form the basement for marine sediments (the breakup unconformity). Erosion of the land in front of the Great Escarpment provided the sediment for offshore deposition, so the volume of eroded sediment should roughly equal the volume deposited offshore. (Source: C.D. Ollier). (b) Diagram of the offshore sediments between Eden and Newcastle by Davies (1975), with calculated volumes of sediment. The larger volume to the south (left) results from erosion of a large area of weathered granite. E = Eden; S = Sydney. (Source: from Davis, P.J. *J. geol. Soc. Aust. 22,* 1975).

The Murray Basin in Australia came into existence at the start of the Tertiary. Before that, drainage from a divide in Victoria, roughly like the present drainage divide, drained north to the Great Artesian Basin. Subsidence caused the drainage to be diverted to the west, and the ancestors of the present west-draining rivers were created.

The point of this account of stratigraphic deposition is that the story of landscape evolution and the history of individual rivers, can sometimes be traced through a long time interval, relating geomorphology to Mesozoic and Cainozoic sedimentary basins.

Since the formation of a new coastline in eastern Australia, the margin of the Australian continent has been eroding, and sediment has been accumulating on the new continental shelf, above what is called the breakup unconformity. This situation presents an opportunity to calculate the volume eroded and see if it matches the volume deposited (Ollier, 1992). This part of Australia has a Great Escarpment (described on p.119), which makes calculation easier. At the simplest the eroded material wedge on land would have the same volume as the sedimentary wedge (Figure 3.8). One place where this calculation is possible is the area between Newcastle and Eden in New South Wales. This is a straight stretch of coastline where a series of offshore profiles have been determined that enable calculation of the sedimentary volume. The volume of material eroded from the land was calculated from a triangular prism with length 550 km, width 50 km (the rough distance between the Great Escarpment and the coast), and height 0.5 km (average height of the escarpment). This is divided by two to allow for the uneroded ridges below the escarpment. The volume works out at 3400 km³. The volume of offshore sediments was calculated by Davies (1975) who divided the area into two as shown in Figure 3.8, obtaining a total of 3400 km³. The coincidence is incredible, but the match suggests the model may be roughly correct.

The Mediterranean Basin records a unique event, known as the Mesinnian Event, which is the drying-out of the Mediterranean about 7 million years ago. This had great consequences on the geomorphology of the surrounding area, but is recorded on the floor of the Mediterranean itself by thick evaporitic sediments (Hsu, 1974) and up to 3 km of Pliocene sediments. The Messinian Event is described further on p.130 (Chapter 9).

Chapter 4
Rivers and valleys

Depositional features of rivers and valleys have already been discussed. In this section we shall look more at drainage patterns and their meaning.

Drainage patterns

Simple river systems have a dendritic pattern with tributaries joining a main stream at acute angles. Any gross departure from this arrangement suggests that the pattern has been affected by some complicating factor of structure, tectonics or geomorphic history. The dendritic pattern is not confined to alluvial plains, but will also be present on well-established plains cut across bedrock. It may even be the pattern for very large-scale reconstructed drainage, like the Tertiary drainage that once flowed across much of Canada towards a sedimentary basin between Canada and Greenland (Figure 4.1). Many people examining drainage pattern are bedazzled by structural control, and see the obvious lineaments if they are present. Lineaments are easy to see, but it is more important to pick out those aspects of the drainage pattern that are *not* affected by structure, and so see the major geomorphic aspects of the drainage pattern. Drainage patterns controlled by structure are described in many textbooks. Here we shall concentrate on those patterns that relate to the age of drainage and associated features.

Antecedent drainage

Antecedent drainage is the term used for a drainage system which has maintained its general direction across an area of localized uplift. An antecedent river is therefore older than the uplift of some high ground that it

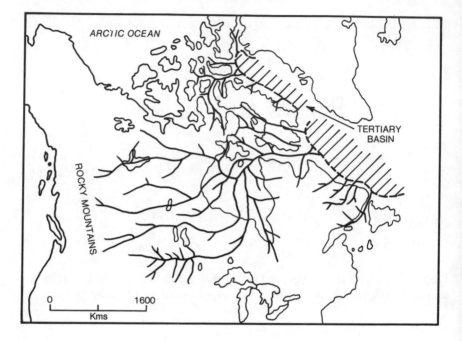

Figure 4.1 Reconstruction of the late Paleocene to Pliocene river system of eastern and central Canada (after McMillan, 1973).

flows across. The river was in its present position before uplift, and erosion has been able to keep pace with tectonic movement.

On a simple tilt block rivers might be expected to flow in the direction of the regional ground surface, but an antecedent river will maintain its earlier course, even if it is in the opposite direction. A splendid example comes from the area south of Milne Bay, Papua New Guinea (Figure 4.2). Here the regional ground surface slopes gently up from south to north, and then descends quickly: it is an obvious tectonic tilt block. The main rivers rise at an elevation of only 50 m near the south coast, but flow north in gorges cut through a range 1000 m high. The rivers evidently originated as north-flowing rivers when the general ground surface also sloped to the north. The rivers continued to flow north while the tilt block was uplifted, and erosion kept pace with uplift so the rivers now flow through gorges cut through the tilt block (Pain and Ollier, 1984). The high part of the ridge includes some remnants of a Plio-Pleistocene erosion surface, so uplift and antecedence are Quaternary events, though the original drainage may date back to the Upper Tertiary.

In China the Fen River has a strange course, almost parallel to the Yellow River (Huang Ho), but it flows across a whole series of horsts and grabens, repeatedly leaving the lowland which would provide an easy passage to the Yellow River and plunging into gorges that cross the uplifted blocks.

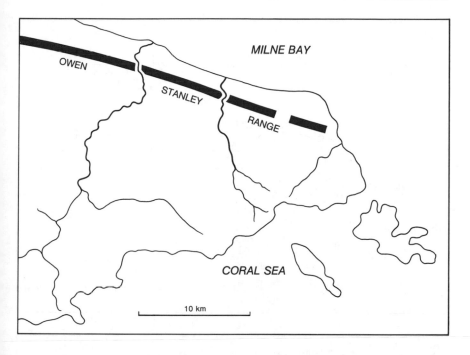

Figure 4.2 Drainage of Milne Bay area, Papua New Guinea. The main river rises within
1 km of the south coast at an elevation below 80 m, but flows north through a deep gorge
across mountain ranges over 1000m high. Other rivers show the same kind of pattern.
(Simplified from Pain and Ollier (1984).

Ahnert (1989) pointed out that the four large rivers of Germany, the Rhine,
Weser, Elbe and Danube, have little relation to the five major landform
regions into which the country can be divided. Only the Danube flows for any
distance along a major regional boundary. The Rhine flows from the Swiss
Alps through all five regions to the North Sea. All the rivers are older than the
uplift of the blocks they cross so that they flow for long stretches in antecedent
courses. The Rhine Gorge between Bingen and Bonn is perhaps the most
spectacular of these antecedent courses.

The development of the Rhine valley through the Rhenish Slate Mountains
was probably initiated during the regression of the sea in the Upper Miocene.
The drainage followed the linear tectonic depressions that were formed in the
Oligocene and Lower Miocene. Superposition of streams by erosion into local
Oligocene gravel fills has also taken place in some valleys (Andres, 1989).

The supreme example of antecedence is provided by some rivers that cross
the Himalayas (Wager, 1937). These mountains form the highest topographic
ridge on earth, and one might expect that rivers would drain away in opposite
directions from the axis of uplift. In fact rivers that rise on the Tibetan Plateau
north of the Himalayas, such as the Arun, flow south through gorges cut across

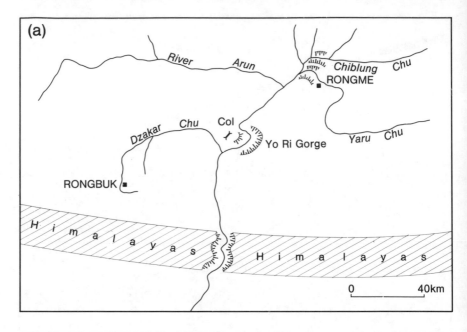

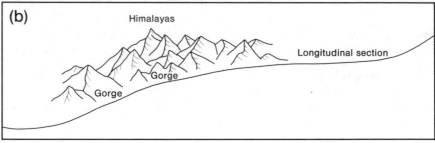

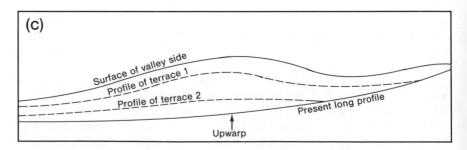

Figure 4.3 The Arun River crossing the Himalayas.
(a) Location map.
(b) Long profile of the Arun River, with the Himalayan peaks superimposed.
(c) Terraces of the Arun River, showing the effect of warping. (Simplified from Wager, 1937).

the Himalayas (Figure 4.3). They are antecedent rivers that flowed south before the Himalayas were raised across their course. That the process has been going on for some time is shown by the warping of old river terraces associated with the rivers. They are bent into gentle curves, with greatest separation from the present river close to the axis of maximum uplift. They also show that uplift is by vertical warping, and not by faulting. There is no direct way of knowing when the original southerly drainage of the Arun was established, but it is post-Cretaceous because gently folded Cretaceous strata lie on the Tibetan Plateau.

Superimposed drainage

Suppose a plain is eroded across varied bedrock, and then covered by a veneer or a thick layer of younger sediment to make a depositional plain. Suppose further that a river flows across the plain and then, perhaps after tectonic uplift, vertical erosion increases. The river may maintain its course, even if hard rock is encountered at some point, and may cut a gorge through the hard rock. This is called superimposed drainage. In areas of superimposed drainage there is often a marked contrast between the pattern of main rivers, which have inherited superimposed courses unrelated to structure, and tributary valleys which may be strongly structurally controlled.

The age of superimposed streams can sometimes be determined from the age of the covering sediment. The River Avon at Bristol flows across a broad clay plain, and then it leaves the apparently easy route to the sea and enters a gorge cut through a plateau of Carboniferous Limestone. The limestone is folded and faulted but has been planated, and an earlier Avon flowed across this plain. Remnants of Pliocene deposits suggest that the plain, and the river, were already in existence in Pliocene times. In the Cape York Peninsula of northern Queensland, Australia, rivers such as the Pascoe are superimposed on to Palaeozoic bedrock from a Cretaceous cover.

Classic examples of superimposed drainage are found in the Appalachian Mountains. Accordant levels of ridge tops show that the folded Palaeozoic strata were planated, and major rivers such as the Susquehana originally flowed across this plain regardless of structure. They have later been incised and now flow through superimposed gorges, but tributaries are strongly structurally controlled. The age of the old planation is controversial, as is the former existence of a Cretaceous cover. The lack of any remnants of a Cretaceous cover makes the idea questionable. It seems equally if not more probable that the area was a Cretaceous peneplain bounding the Cretaceous transgression to the east. The term 'superimposed' can still be applied, even though the drainage is superimposed only from an old plain and not from a thick sedimentary cover. Thornbury (1969) suggests that the drainage may have been along present lines since Triassic times. The streams would then have been draining the Appalachian region to the line of rifts and lakes that were a Triassic precursor of the opening of the Atlantic. Meyerhoff and Olmsted (1963) believe that the present drainage lines are direct descendants of Permian streams. They stress the supposed coincidence between present stream courses through ridges and structural sags and fault zones, a situation unlikely to occur if the drainage lines are superimposed from a Cretaceous cover.

Radial drainage

Radial drainage is common on volcanic cones. This is an unremarkable feature, unless the volcano is in a stage of advanced erosion, when the radial drainage, superimposed on to sub-volcanic bedrock, may indicate the former extent of the volcano. The 19 million-year-old Ebor Volcano of New South Wales has a radial pattern (focused on a gabbro plug) that persists where the volcanic rock has been almost entirely removed, and the changes in direction of the rivers probably indicate the former extent of the volcano (Figure 4.4).

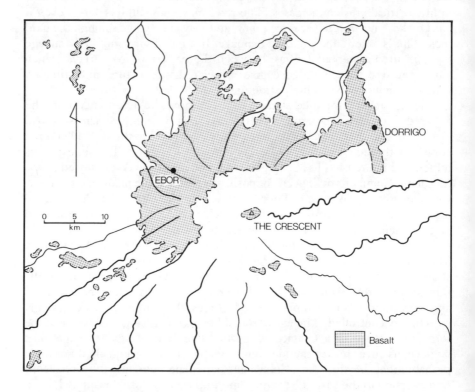

Figure 4.4 Radial drainage of the 19 million year-old Ebor Volcano, New South Wales, largely superimposed on to Palaeozoic bedrock. (C.D. Ollier).

The 54 million-year-old Monaro Volcano is in a greater state of erosion, but again the radial drainage indicates something of its former extent. River captures in the sub-volcanic rocks have complicated the picture, but the pattern is still clear, and quite unrelated to the structures of the sub-volcanic bedrock (Figure 10.10).

Domes caused by broad uplift may also develop radial drainage. A classic example is the Lake District of northern England (Figure 4.5). Most authorities suggest that there was a Cretaceous cover, or at least a Cretaceous erosion surface before domal uplift, and the present drainage is superimposed from that dome. Certainly the radial drainage pattern has nothing to do with the complex bedrock structures of the region. The basic features of the drainage thus date back to the Cretaceous. A partial radial pattern is discernible in the reconstructed ancient drainage of Wales (Figure 4.6), as deciphered by Brown (1960).

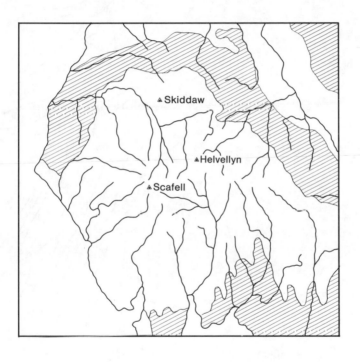

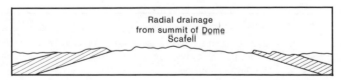

Figure 4.5 The dome of the Lake District, northern England. (a) Radial drainage, and very simplified geology. The shaded area represents Carboniferous Limestone. (b) Cross section of the dome.

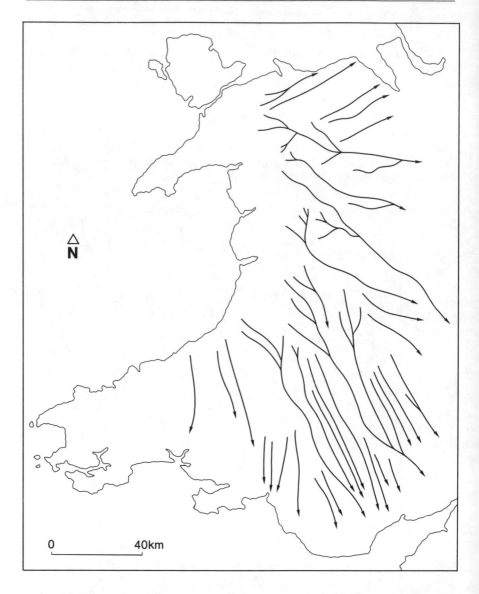

Figure 4.6 The original drainage pattern of Wales, as interpreted by Brown (1960). It is a partial radial system, centred on a high point in northern Wales. (Based on E.H. Brown, *The relief and drainage of Wales*, (1960).

Figure 4.7 shows the radial drainage associated with the Dundas Dome in western Victoria, which is probably of late Tertiary age. Besides the radial drainage this example has a river that flows around the perimeter of the dome in an annular pattern, the Glenelg River, which, like the radial drainage, must

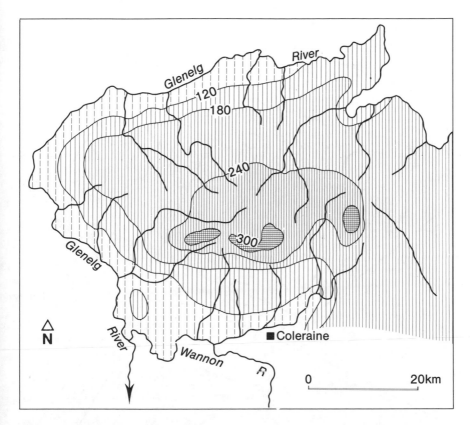

Figure 4.7 Radial draining on the Dundas Dome, Victoria, and annular drainage of the Glenelg River.

post-date the formation of the dome. The Brazilian Shield described later can be regarded as a mega-dome if the hypothesis of Grabert is correct.

River capture

River capture is the name given to change in river pattern produced when a stream which is cutting down rapidly extends headwards until it reaches a slowly eroding river, which is then diverted into the capturing stream. In Figure 4.8 stream CD is steep and is eroding downwards and headwards quite quickly. Stream AB has a gentle gradient and is cutting down only slowly. Stream CD eventually captures stream AB, creating several features. Beyond the point of capture, marked x, the stream xA is much reduced in volume compared with the old stream AB, but the volume of water going down river CD is now augmented by the extra water gained from section xB, which enhances its erosive power even further. The sharp change in direction of the river at x is called an elbow of capture, and the point where the gradient

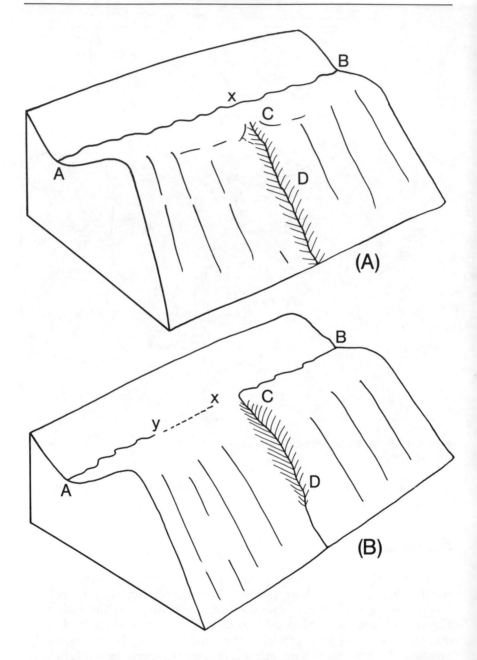

Figure 4.8 Diagram of river capture. For explanation see the text. (Source: C.D. Ollier, *Tectonics and Landforms*, Longman, 1980).

suddenly increases is called a knick point. The knick point will in time migrate upstream towards B. The section of the old valley AB between x and y no longer carries water, and is called a wind gap. Old river sediments may be preserved here, and are often the best indication of the age of river capture.

A neat example of river capture is provided by the Barron River near Cairns, Queensland (Figure 4.9). The upper course of the river is on a plateau, but it turns seaward and drops 300 m over large waterfalls and crosses a lowland plain to the Pacific. The headwaters were originally tributary to the Mitchell, which even now rises within 15 km of the Pacific but flows 550 km to the Gulf of Carpentaria. The wind gap contains a continuous belt of river gravels marking the old river course. Unfortunately in this instance the gravels cannot be dated and the age of the capture is not known. Similar captures are common in Queensland, and they are probably a few million years old.

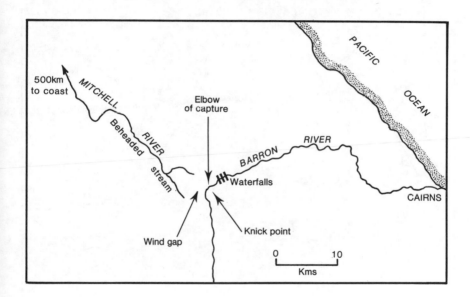

Figure 4.9 River capture of the former headwaters of the Mitchell River by the Barron River, Queensland.

Reversed rivers

The normal pattern of rivers is dendritic, with tributaries joining a main stream at acute angles that point downstream. The tributaries also have steeper gradients than the main stream. If the whole area is tilted back, the main stream can be reversed, that is, it flows in the opposite direction to its old course. The tributaries, being steeper, are not affected by the tilting and continue to flow in the same direction. The tributaries are then said to be barbed: they still join the main stream at acute angles, but the angles point upstream relative to the main stream.

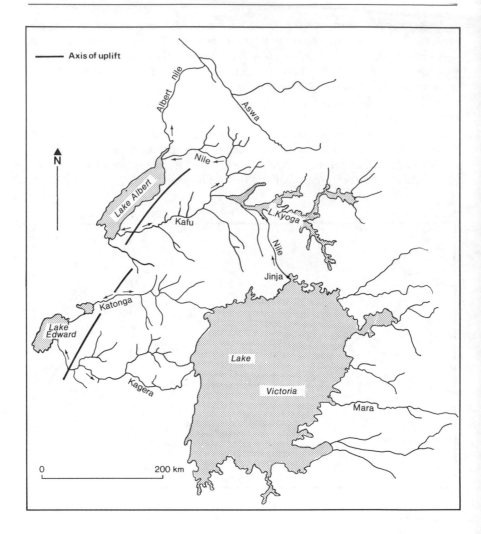

Figure 4.10 Reversal of the Kafu-Kagera-Lake Kyoga system. The original drainage was from east to west, with the Mara continuous with the Kagera. The drainage lines of the Katonga and Kagera are continuous between Lake Victoria and Lake Edward, as is the Kafu from Lake Kyoga to Lake Albert. Much of the drainage of these rivers is now reserved as shown by the arrows because of uplift in a zone parallel to the rift valley faults. The middle course of the Mara–Kagera river was drowned when downwarping created the Lake Victoria basin. Lake Victoria overflowed at the lowest point on its watershed at Jinja to form the stretch of the Nile between Lakes Victoria and Kyoga. Lake Kyoga was formed by similar back tilting of the Kafu River. It flowed up a northern tributary and overflowed into the Albert rift valley, forming the stretch of the Nile between Kyoga and Albert. From Lake Albert the Nile flows north along a continuation of the rift until it joins the Aswa, which follows a major Precambrian mylonite band. The disruption of drainage and the formation of Lake Victoria dates backs to the Oligocene. (Source: C.D. Ollier, *Tectonics and Landforms*, Longman, 1980).

The Kafu River in Uganda provides a good example (Figure 4.10.) It used to flow from east to west, but has been reversed by relative uplift of the shoulder of the Lake Albert Rift Valley, and now flows to the east, with barbed tributaries maintaining their old direction. A small stretch of the Kafu beyond the axis of uplift retains the old direction, and there is a broad swampy valley between the old course and the headwaters of the reversed course. The rift valley was initiated by movements in the Oligocene and reversal may date from then, and certainly the original course of the valley must be at least that old, and possibly Mesozoic.

Another good example is provided by the Clarence River in northern New South Wales. This flows along the edge of a Jurassic sedimentary basin, which was in fact deposited by some ancestor of the Clarence and similar streams coming from the east, where the Pacific Ocean now lies. The river is partly superimposed, presumably from a more extensive Jurassic cover. The most obvious feature is the barbed drainage on the southern side (Figure 4.11). (The northern side is ignored here because the drainage pattern is complicated by volcanic activity.) The headwaters (known as the Orara River) flow normally with simple dendritic drainage, and beyond the Great Divide the Condamine, which appears to be a former continuation of the Clarence, also has simple dendritic drainage. But between the Great Divide and the Orara junction all tributaries are barbed, suggesting reversal. This fits in with coastward tilting associated with the tectonics of the continental margin. The time of reversal is not known with precision, but is probably Upper Cretaceous or Early Tertiary.

It must be pointed out that barbed drainage can result from events other than reversal, such as river capture of streams flowing in an opposite direction. A broad range of landscape features should be included in deciphering river evolution and not the pattern alone.

A different approach to reversal is presented by Grabert (1971). He believed that the Brazilian Shield, with its thick weathering cover, had been exposed since the Triassic or earlier, and from it a radial drainage flowed toward surrounding seas. At that time there was no Atlantic to flow to, nor any Andes mountains. The west-flowing rivers from the Shield would have flowed to the Andean depositional trough or geosyncline. But the Mid-Tertiary uplift of the Andes reversed the pattern and caused the Amazon to flow eastwards (along a structural low) and the Parana to flow southwards subparallel to the strike of the Andes. An alternative explanation for the Parana is by Cox (1989) who believes there is a hot-spot dome in east Brazil, around which the Parana would have to flow. This simple hypothesis does not include as much geological and geomorphic detail as that of Grabert.

Exotic terranes and drainage patterns

The concept of exotic terranes is described in Chapter 8. These are fragments of the earth's crust, sometimes actual islands, that collide and amalgamate. If

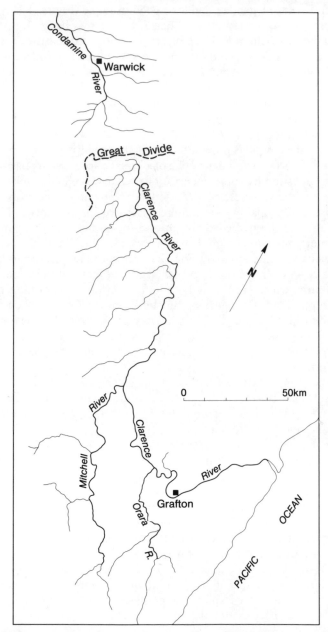

Figure 4.11 Barbed tributaries of the Clarence River, New South Wales. Only southern tributaries are shown as the northern side is complicated by volcanic derangement of drainage. Beyond the Great Divide the Condamine River was a continuation of the drainage before reversal of the Clarence. The Orara in the south, and the parallel tributary, the Mitchell, still has its original northerly direction. The course of the Clarence between the normal drainage of the Orara and the reversed Clarence flows along a tectonic trough to the sea (after Haworth and Ollier, 1992).

two terrestrial terranes, each with its own drainage pattern, should collide, then two completely different drainage patterns would be brought together.

One possible example comes from Papua New Guinea (Ollier and Pain, 1988). North-flowing rivers to the south of Milne Bay were shown to be antecedent to tilt block movement that raised mountains about 1000 m. The extension of these rivers to the north of Milne Bay could not be traced, which may be because the land to the north is an exotic terrane as described by Pigram and Davies (1987).

The Sepik–Ramu drainage in Papua New Guinea, on the terrane hypothesis, follows the suture between different terranes (Figure 4.12). If the exotic terrane hypothesis is correct, and as widespread as its proponents suggest, there should be many more such rivers following sutures, but so far work on terranes has largely ignored the obvious landforms and focused on remote geophysical data.

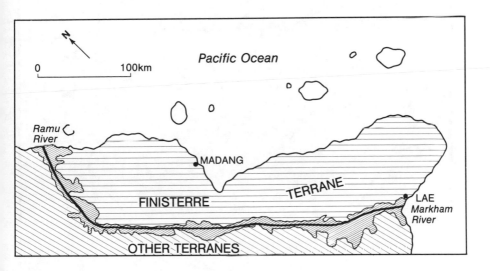

Figure 4.12 The Sepik-Ramu drainage of Papua New Guinea, which possibly follows the suture (shown by the bold line) between exotic terranes.

Another example is from the Kimberley Plateau of north-west Australia. The Plateau consists of Proterozoic rocks, and the planation surface which bevels it existed in late Proterozoic times, about 700 million years ago. As explained on p.24, there is reason to suppose that most of the valleys were in existence at the time of Precambrian glaciation. They were first filled with glacial sediments, and then exhumed from a cover of Precambrian sediments. The significant feature is the drainage boundary (Figure 4.13). This has been called the Kimberley Rim, and separated almost completely the drainage of the Kimberley Plateau from that outside. Only one river, the Fitzroy, crosses the Kimberley Rim, and there is a good structural explanation for that. A possible explanation for the drainage pattern is that the Kimberley Block is an exotic terrane that docked against Australia, but retained its independent drainage pattern.

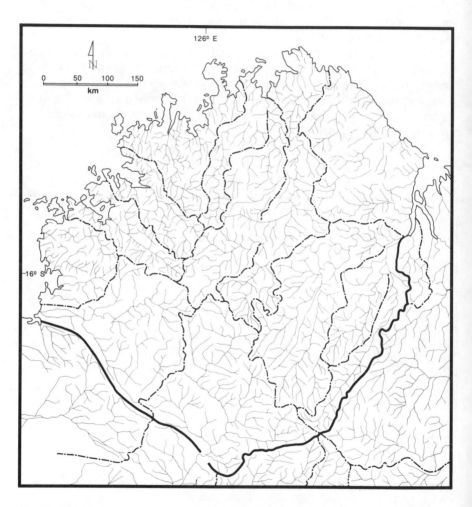

Figure 4.13 Drainage pattern of the Kimberley Plateau and surroundings. The thick line marks the Kimberley Rim, which virtually isolates the drainage of the Kimberley terrane from the rest. Only one river crosses the Kimberley Rim, and this follows a structural weakness.

Possibly related to exotic terranes is the strange pattern in central Italy (Figure 4.14). The Adriatic slope is characterized by parallel drainage, with closely spaced watercourses heading almost directly to the sea, flowing across folded rocks with little regard for structure. The drainage to the Tyrrhenian

Sea has a reticulate pattern, largely controlled by extensional faults. Although these marked differences suggest the watershed might be a terrane boundary, Mazzanti and Trevisan (1978) believe the Apennine divide has migrated north-eastwards from the Upper Miocene to the present, and marks the boundary between tectonic extension and compression.

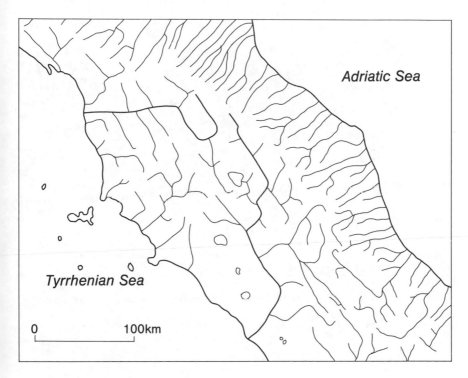

Figure 4.14 The drainage of central Italy, showing the contrast between drainage to the Adriatic and that to the Tyrrhenian Sea. Possible explanations are given in the text. (Simplified from Mazzanti and Trevisan, 1978).

Drowned drainage patterns

The present is a time of high sea-level, and for much of the past rivers had longer courses across what are now continental shelves. For example, a major river system that functioned at times of low sea-level can still be traced on the Sunda Shelf (Figure 4.15). Freshwater fish of west Borneo are more like those of east Sumatra than those of east Borneo, suggesting contact at that time. Similarly, freshwater fish did not cross from Kalimantan to Sulawesi, suggesting a marine barrier between the two, even though Palaeomagnetic evidence suggests west Sulawesi has been close to Kalimantan since Cretaceous times. Similar offshore rivers have been traced elsewhere in the

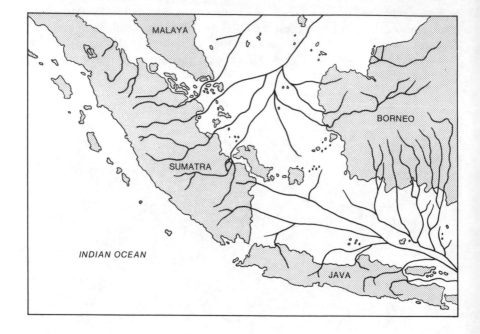

Figure 4.15 Part of the river pattern on the Sunda Shelf at times of low sea level. (Source: C.D. Ollier, *Modern Quaternary Research* in *S.E. Asia*, 9, 25-42, 1985).

world, such as the extension of the Rhine–Thames drainage system into the North Sea.

Ancient big rivers

Some of the world's big rivers can be traced back into geological time. For instance, the drainage of Uganda that was reversed to form Lake Victoria and Kyoga (p. 40) formerly drained into the Congo (Figure 4.16). These Congo headwaters were beheaded by the formation of the Western Rift Valley, and such a major change should be reflected in Congo geomorphology and in the sediments of the Congo delta. The beheading probably occurred in the Miocene.

The beheaded rivers were diverted north as the White Nile, which in fact provides most of the water in the Nile. Other big rivers that were connected to the Nile have now been detected in the Sahara by remote sensing (Burke and Wells, 1989). Further evidence of the antiquity of the Nile is shown by the deep channel it cut in the Messinian Event about 7 million years ago (see p. 13).

Ancient big rivers can be identified by a combination of sediment-type mapping, palaeocurrent study and the careful reconstruction of the tectonic history. Ancient or modern, the location of big river systems on cratons largely

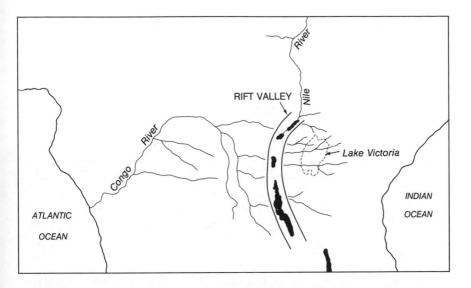

Figure 4.16 The Congo-Nile relationship. Before the formation of the rift valley the drainage of much of Uganda formed the headwaters of the Congo, but rifting and associated complications diverted these rivers to the Nile.

follows structural lows such as deep-seated rifts, aulacogens, and geofracture systems. Big river systems have their greatest longevity on cratons where some have persisted as long as one-sixteenth of earth history (Potter, 1978). A major factor in any river system is the marine history of its drainage basin, a history that in turn depends on the region's tectonic history.

The ancestral Mississippi, called the Michigan River by Swann (1963), was described earlier (Figure 3.5). It has been oriented south and south-westward across the North American craton since the early Carboniferous.

Chapter 5
Weathering

Weathering is the alteration and breakdown of rocks near the earth's surface, mainly by reaction with water and air, to form clay, iron oxides and other weathering products. It is generally assumed to be associated with the present ground surface and to form soils, but there is abundant evidence of deep weathering in which alteration takes place to depths of hundreds of metres. Such weathering takes a long time.

The study of weathering includes observations on changes in materials, such as the changes in mineralogy as a rock weathers, and the formation and destruction of weathering profiles. A typical weathering profile is shown in Figure 5.1. The upper part is usually disturbed by processes such as hillside creep or churning by burrowing animals or tree roots. The lower part is often undisturbed and retains structures of the original bedrock even though the minerals have been extensively changed: it has suffered isovolumetric alteration. This material is called saprolite. The term regolith is used to describe all the unconsolidated surficial materials, including both saprolite, disturbed layers, and even surficial sediments. The base of weathering is seldom even, but usually penetrates along joints and other fissures, isolating corestones of less weathered rock. The junction between weathered and fresh rock is called the weathering front, and is commonly very sharp.

The meaning of 'age' of weathering

In principle the age of weathering cannot be older than the age of the rock that is weathered, nor younger than the age of the overlying unweathered rock. Direct dating of the age of weathering is sometimes possible by techniques such as palaeomagnetism.

As soon as a rock or sediment is exposed to air and water it is liable to alteration by 'weathering'. With the passage of time more minerals may alter,

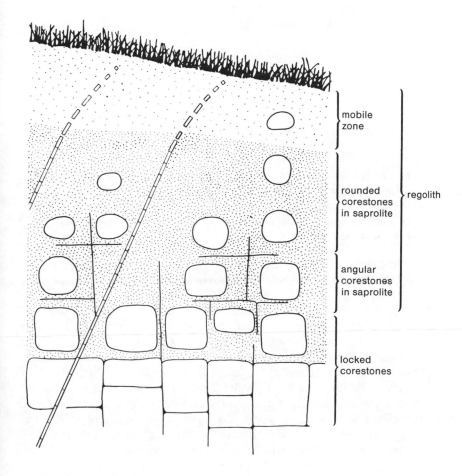

Figure 5.1 A typical weathering profile in granite. Saprolite is weathered rock in place, as indicated by unmoved quartz veins. Alteration is isovolumetric. The upper zone may be moved by hillside creep, burrowing animals, plant roots, etc. to form a mobile or re-sorted zone. Regolith is a term that covers all the unconsolidated material near the earth's surface, including saprolite and even surficial sediments. (Source: C.D. Ollier, *Weathering and landforms*, Macmillan, 1990).

and early weathering products may be altered to more weathered products. A silicate mineral, for example, may be weathered first to a smectite clay mineral, then to a kaolinite clay, and perhaps finally to a bauxite.

A sequence may be imagined for minerals, weathering profiles or soils from the earliest stages of weathering through intermediate stages to the later stages and possibly an ultimate stage. Mohr and van Baren (1954) proposed the following stages of soil formation:

Initial stage (unweathered parent material)
Juvenile stage (much parent material is unweathered)
Virile stage (easily weatherable minerals have decomposed)
Senile stage (only the most resistant minerals have survived)
Final stage (soil development complete, soil is weathered out under the prevailing conditions)

The anthropogenic nomenclature seems to be especially popular with soil writers, who also tend to include horizon formation in their definitions. Millar, Turk and Foth (1966) provide the following list:

Immature soils are characterized by organic matter accumulation at the surface and little weathering, leaching or eluviation.
Mature soils are characterized by the development of a B horizon.
Old-age soils have very marked differences between the properties of the A and B horizons.

Marbut (1928) defined a mature soil as one 'whose profile features are well developed'. This very subjective definition is very hard to use in practice. Jenny (1941) wrote 'Mature soils are in equilibrium with the environment.' This is similar to Nikiforoff's more complex statement (1949) 'A mature soil represents the steady state of the dynamic system comprising the soil and its environment, the latter including the climate and the organic world.' Such definitions do not refer to the weathering profile or soil profile, but emphasize processes, the environment and the balance between factors. On this basis it is impossible to be sure whether a profile is mature or not.

Any given weathering property may change through time, and when change has ceased or is asymptotically approaching zero it may be said to be 'mature'. But not all properties approach maturity at the same rate. If and when *all* properties have reached the point of negligible further change, the weathering profile as a whole may be called 'mature'.

It should be noted that a mature soil is not necessarily inert, with all chemical processes at a standstill. Soil properties may vary with changing environmental conditions, such as temperature and moisture. On the long time-scale environmental conditions have often changed significantly. A weathering profile formed under tropical conditions will be out of equilibrium if it comes to be in a periglacial environment. Many examples of such change are found in the real world.

The concept of an ultimate weathering profile is also difficult to support in theory. One ultimate profile consists almost entirely of iron oxides, hydroxides and kaolin. There is something about the bonding of kaolin and iron oxides that makes this a very tough material, tending to be highly resistant to later change. Nevertheless profiles on such material do seem to have some changes, in organic matter accumulation and reaction, changes in groundwater hydrology, and others. Some writers believe that ferricretes in such profiles may be in some kind of dynamic equilibrium or steady state. Radwanski and Ollier (1959) described an East African soil with mottles and concretions in the subsoil which are never exposed at the surface, and suggested that the process of iron movement and precipitation is taking place all the time: the mottled subsoil which are never exposed at the surface, and suggested that the process of iron movement and precipitation is taking place all the time: the mottled subsoil is gradually destroyed from above, but develops downwards by encroaching on the underlying deeply weathered rock. Tardy and Nahon (1985) have suggested that successive ferricretes (laterites) are in steady state relationship with landscape evolution in parts of West Africa.

Yet some weathered profiles, especially those rich in iron oxides, have been dated by palaeomagnetism to Tertiary and even Mesozoic times. Such materials must indeed be 'fossil' and not in equilibrium with environmental conditions of today. Whatever changes are happening on such materials today they are sufficiently minor for the magnetic signal of the past to be retained.

In other situations, it is clear that environmental conditions of today are leading to erosion of deeply weathered material, exposing fresh rock. Present-day weathering is not responsible for the thick regoliths observed, which were therefore formed in some earlier weathering period which is potentially datable.

In brief, there are two main 'ages' that can be applied to weathering in the landscape. The first is the age of initiation of weathering. The second is the age or time when the profile is well developed under prevailing environmental conditions and can be dated by techniques such as palaeomagnetism and stable isotope studies, or by overlying strata which effectively halted weathering. We shall look at examples where the combination of geology and geomorphology has enabled some age limits to be put on weathering.

Examples of ancient weathering

The plateau deposits of east Devon contain a variety of residual deposits including soft kaolinitic weathering profiles and silcretes (Isaac, 1983). These reflect a complex history that began with the emergence of the post-Chalk land surface at the end of the Cretaceous. During the Paleocene residual flint gravels up to 10 m thick accumulated, and deep weathering profiles of the same age are preserved in irregular pockets in the Chalk. Locally, late silicification of the weathering profile produced silcrete. In some areas the Paleocene weathering profile was destroyed by Middle Eocene times when new kaolinitic weathering profiles were established on freshly exposed Palaeozoic rocks (Bristow, 1968).

Le Coeur (1988) has described weathering from Scotland which could not occur under the present climate. In the absence of any datable material, he attributed it to end-Tertiary times. Rather similar weathering profiles in Scotland, with pre-glacial saprolite over 50 m thick, have been described by Hall (1987) who believed they are of Tertiary age. In the Buchan area of Scotland outliers of Old Red Sandstone (Devonian) resting on Caledonian basement show that erosion was already close to present levels by the Late Palaeozoic (Hall, 1987).

Lowering of relief continued throughout much of the Mesozoic. During the Cainzoic, weathering under warm humid environments produced at least two different generations of weathering cover and a subdued relief (Figure 5.2). Hall wrote, 'The exceptionally long morphogenic history of Buchan reflects post-Palaeozoic tectonic stability.'

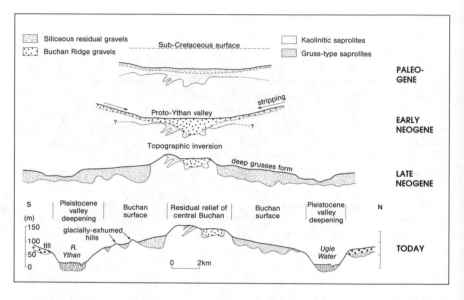

Figure 5.2 Weathering at Buchan, Scotland. (After Hall, 1987)

From southern Sweden, Lidmar-Bergstrom (1988) has described deep weathering profiles exposed in a kaolin pit, in which corestones of granite are exposed (Figure 5.3). Cretaceous marine sediments overlie the weathered material, and there are even Cretaceous oysters attached to some of the corestones. The weathering must therefore be at least as old as Cretaceous. This is especially significant because it is often assumed that Quaternary glaciation removed all old regolith from much of the northern hemisphere, but we now know that pockets of Mesozoic weathering have been preserved. The ice sheets may have merely trimmed landscapes rather than totally reshaped them.

Iron-rich regolith materials in southern Germany are found on the highest planation surfaces, and are of Early Tertiary or even Cretaceous age. A pre-Upper Cretaceous weathering horizon occurs in north-east Bavaria, consisting of thick iron ore developed by weathering *in situ*, which was covered by younger sediments (Bremer, 1989).

The ancient (Eocene) valleys of the Kalgoorlie district have already been described (Figure 3.2). Sediments filling these valleys have been weathered, but not sufficiently to destroy the organic remains preserved in coal deposits at the base of the sediment. However, the valleys are cut into a deeply weathered profile which can be divided into an oxidized upper zone and a

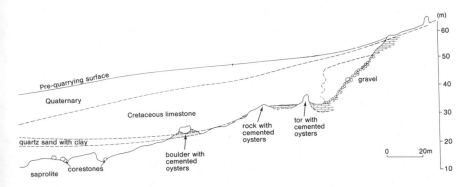

Figure 5.3 Section through a kaolin quarry in Sweden. Deep weathering of granite with corestone formation was followed by stripping. The irregular surface was then covered by marine sediments, of Cretaceous age as indicated by fossils including oysters growing on corestones. There was renewed natural erosion, and modern quarrying. (After Lidmar Bergstrom, 1988).

reduced lower zone. The oxidized zone is so intensely altered that it seems most unlikely that fossil fruit, leaves and pollen could survive such weathering. It seems much more probable that the valleys were eroded into an already deeply weathered plain. Since the valley fill is about 40 million years old, the weathering of the valley sides must be much older still. Indeed, if the valleys pre-date continental drift, as explained on p. 19, they must be older than 55 million years (when Australia started drifting away from Antarctica) and possibly older than 100 million years (when the first rifting occurred). This weathering could therefore be older than 100 million years, perhaps Lower Mesozoic, perhaps even older. Alunite in the weathered rocks from central Western Australia gave a potassium/argon date of about 60 million years, which is regarded as a minimum (Bird et al., 1990).

In the precious opal fields of South Australia, alunite is commonly associated with opal and gypsum, in Cretaceous host rocks. Potassium/argon ages suggest several episodes of opal formation during the Miocene (Bird et al., 1990).

Ball clay is a kaolinitic sedimentary clay deposited under freshwater conditions and frequently associated with a chemical weathering profile formed at the same time as or before the deposition of the ball clays. A survey of ball clay occurrences in Western Europe and North America (Bristow, 1990) showed that they are nearly all of a similar age — Eocene to Lower Oligocene. This widespread development of ball clay at an Eocene palaeolatitude of 40–45 degrees N seems to coincide with a period of high palaeotemperature in the Eocene, when special climatic conditions developed which are not present anywhere on earth today.

Commercial sedimentary kaolins in Georgia, United States, are Late Cretaceous to Middle Eocene in age. According to Pickering and Hurst (1989), they originated during a time of pronounced global submergence and warmer climates, when most of the world's kaolin and bauxite deposits formed. The ultimate source of the kaolinitic sediments was deeply weathered rocks in the Piedmont Upland, where low relief, extensive outcrops of granitic rocks, and a humid subtropical to tropical climate caused the development of extensive thick saprolite, some containing more than 60 per cent kaolinite. Cretaceous kaolins appear to be partly fluvial and partly deltaic, and commonly they have been exposed to post-depositional weathering and diagenesis. The Tertiary kaolins, in contrast, are nearshore marine deposits little affected by later weathering or diagenesis. A pre-Cretaceous laterite has been reported from New York (Blanck, 1978), and deep kaolinization of bedrock at Boston of probable Mesozoic age is reported by Kaye (1967).

In many areas of lava flows, deeply weathered layers, sometimes called bole, are found between lava flows. These have sometimes been dated by fossils, sometimes by palaeomagnetism. Many turn out to be Tertiary or Mesozoic in age. They become significant in present-day landscapes when they outcrop at the ground surface. In parts of New South Wales, for example, they tend to form structural terraces as overlying lava flows retreat faster than the bole. It would, of course, be a great mistake to think the observed weathering profiles on such terraces are in any way related to the present climate or geomorphic position. Palaeomagnetic data from Australian basaltic regolith have been described by Schmidt and Ollier (1988). Tertiary weathering is common, and some goes back to the Cretaceous.

Sub-basaltic weathering

The effect of a basalt flow on the weathering of underlying sedimentary rocks has been described by Schmidt et al. (1976). It might be expected that a basalt flow would protect underlying material from weathering and that neighbouring rocks with no basalt cover, exposed to the atmosphere, would weather faster. In fact the reverse happens, and weathering is much more intense below the flow (Figure 5.4). Two possibilities can be suggested to account for the distribution of weathered rock beneath the basalt:

1. There has been an increase in the weathering of Ordovician rocks beneath the basalt.
2. The weathering of the Ordovician rocks was most intense in low, wet areas that were subsequently covered by basalt.

The junction between weathered and fresh bedrock is sharp, and dips steeply from the original position of the basalt edge, at a gradient that is not parallel to an earlier valley side or to any rock structure. Since the weathered rock is found under all areas ever covered by basalt, and is absent from all valley sides that never had a basalt cover, it seems most probable that the intense weathering is post-basalt. The age of the weathering is younger than the

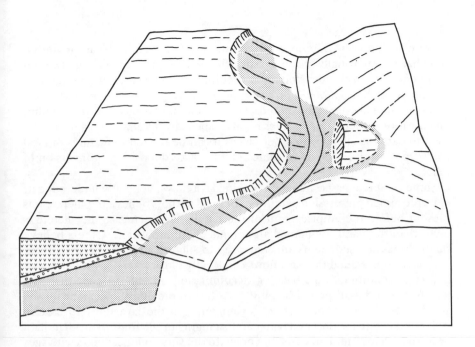

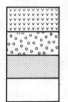

Basalt

Sub-basaltic sediment

Sub-basaltic weathered rock

Fresh Bedrock

Figure 5.4 Sub-basaltic weathering at Eppalick, Victoria. Weathered shale extends to the original edge of the late Pliocene lava flow. (Re-drawn from C.D. Ollier, *Volcanoes*, Blackwell, 1988).

overlying lava flow, an exception to the law of superposition. In South Australia Schmidt found Tertiary weathering (indicated by palaeomagnetism) under a Jurassic lava flow (Schmidt et al., 1976).

Duricrusts

A number of hard layers found in soils and surficial deposits are known together as duricrusts, and separately by names that indicate their main component — ferricrete, silcrete, calcrete and others. The aluminium crust is seldom called alcrete and is widely known as bauxite. The term 'laterite' includes ferricrete but is also used to describe soils, red-weathering and weathering profiles, and there is much disagreement on its definition.

Silcrete

Silcrete is a greyish white rock (sometimes known in Australia as greybilly), very hard, and usually made by silicification of pre-existing quartz sand sediments. It can also develop in fine sediments, in weathering profiles, and in more diverse situations. The silica content is very high, and often the only other components are the noted residual elements zircon and titanium. Silicification sometimes preserves fossils, especially fossil wood, which can be used to date the silcrete. In other places it can be related to features of local landscape history, such as volcanic activity, lake deposition or inversion of relief.

Some silcrete is associated with basaltic lava flows, generally underlying it, or its presumed pre-erosion extension. Several authors believe the silcrete was present before the emplacement of the basalt, in which case the silcrete is older than the basalt by normal arguments of superposition. Others believe that sub-basaltic sediments are converted to silcrete after the valley is filled with lava. It is argued that lava flows can locally block rivers, giving impeded drainage conditions suitable for accumulation of silica, and that the weathering of basalt provides abundant silica to provide the cement for the silcrete. In this case the silcrete is younger than the lava flow. In eastern Australia silcrete is associated with basalts ranging in age from about 40 million years to perhaps 2 million years, but not with the very youngest flows with ages in tens of thousands of years.

Silcrete is also found in other localities. In semi-arid inland Australia silcrete is associated with deep weathering profiles on Cretaceous sediments (Figure 7.3), but its precise age is not known. In the Shoalhaven valley in New South Wales alluvium of Lower Tertiary age has been converted to silcrete (Nott, 1990). At Lake George, an area of internal drainage near Canberra, silcrete has formed near the ground surface in sediments that must be Upper Quaternary or Recent.

Silcretes in South Africa appear to be of Tertiary age, and siliceous materials equated with silcretes in England, France and the United States are also Tertiary features. Silcrete in Devon, England, was formed in the Paleocene (Isaac, 1983). Molina et al. (1990) describe a weathering profile near Salamanca in western Spain in kaolinized granite overlain by fluvial sediments about 58 million years old (Eocene). Both the fluvial series and the upper part of the weathered granite are silicified by opal.

Ferricrete

Hard, iron-rich materials in sediments and weathering profiles are often called ferricrete. Iron may be added to pre-existing material (absolute accumulation), or may be concentrated by the removal of other materials (relative accumulation). Iron is a mobile element if in solution as ferrous iron, and in combination with some organic compounds (chelation), but once it is oxidized it is usually fixed. Many hypotheses of ferricrete formation are based on moving iron in solution to sites where it is irreversibly precipitated.

Some hypotheses of ferricrete formation are based on its association with underlying weathered materials which are depleted in iron. Since the lower material is iron-poor and the upper material iron-rich it seems reasonable to assume that the iron has somehow moved upwards in solution, aided by groundwater fluctuation, capillary action or other possible methods.

Another series of hypotheses suggest that lateral movement of iron is important, and that it is mostly precipitated on lower slopes and in valley bottoms. If this is followed by inversion of relief the familiar situation is produced with ferricrete on plateau tops (Figure 5.5). The distribution of

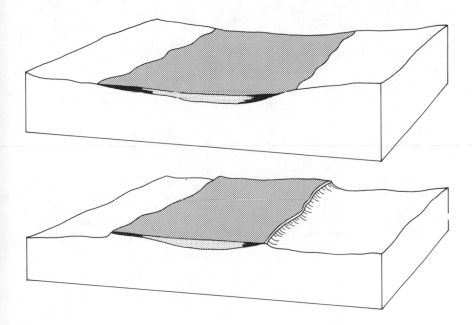

Figure 5.5 Top: Precipitation of ferricrete on lower slopes and valley bottoms. Bottom: Inversion of relief, to produce a ferricrete-capped mesa.

ferricrete in plan may resemble a drainage pattern (Figure 5.6), and there may be associated alluvial gravels. In this case the ferricrete can be tied into the local geomorphic history. For instance, in the Kalgoorlie area of Western Australia, ferricretes are on plateaux separated by broad valleys, and the valleys contain Eocene sediments. The ferricrete is therefore of Eocene age or older (Ollier et al., 1988a). The concept of ferricrete formation in valleys, followed by inversion of relief to locate the ferricrete on hilltops was extended to many other parts of the world (Ollier and Galloway, 1990). It has also been suggested that repeated formation of ferricrete and inversion of relief is possible, a mechanism invoked in Madagascar (Maignien, 1966), and

Australia (Ollier et al., 1988a). This is a widespread method of landscape evolution to be contrasted with the models of Davis or Penck.

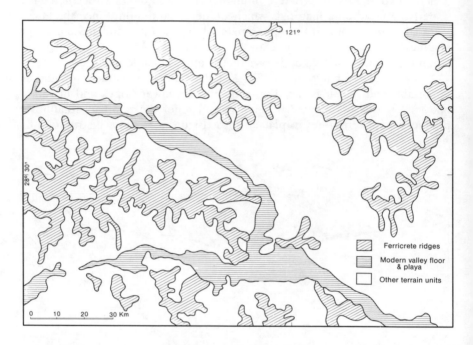

Figure 5.6 Ferricreted ridges in plan, near Kalgoorlie, Western Australia. The shape is like a dendritic drainage pattern, suggesting the ferricrete was formed in ancient valleys and then put on hilltops by inversion of relief. The modern valleys are of Eocene age at least, so the ferricrete is probably older.

Ferricretes and associated deposits can also be dated by palaeomagnetism. Nodular and concretionary ferricrete is usually unsuitable, but underlying red clays often give good results. Many ferricretes have been dated, yielding ages that span the Tertiary.

The literature is full of references to 'laterite' which commonly refers to ferricrete, but may also refer to associated weathering profile, to red soil, or dubious materials of red colour. The word 'laterite' strictly refers to the mottled zone of a weathering profile in saprolite, which is hard enough to be cut into bricks for building houses, and even cathedrals and forts (Ollier and Rajaguru, 1989).

Milnes et al. (1985) made a significant contribution to resolving the 'laterite' problem in their work in South Australia, where terrestrial landscapes have existed since the Carboniferous to Permian glaciation. Widespread weathering profiles and ferricretes on present highland surfaces had been attributed by earlier workers to Tertiary weathering, but detailed examination showed reworking and continuous weathering of relic landscapes since early Mesozoic times, leading to the intricate pattern of sediments and soils forming the present regolith. Sporadic ferricrete crusts are interpreted as remnants of iron-cemented sediments of ancient valleys. The thick kaolinized bedrock beneath the highland surfaces, regarded previously as the mottled and pallid zones of a 'laterite' profile, is the integrated product of weathering throughout the Mesozoic and Cainozoic, and cannot be assigned to distinct climatic events.

However, some profiles can be related to particular events. Idnurm and Senior (1978) described two widespread iron-enriched weathering profiles from Queensland. One is Late Cretaceous to Early Eocene, and the other is of Late Oligocene age. These ages are consistent with times of high palaeotemperature indicated by oxygen isotope analysis of marine fossils in the Indian Ocean. Normal and reversed polarities were obtained from each profile, indicating that the weathering events spanned at least one reversal of the geomagnetic field. The minimum duration of weathering for the older profile is estimated at 10 000 years.

Schmidt et al. (1983) determined the age of laterites in India. The high altitude laterites have undergone a complex magnetic history during the Late Cretaceous and Early Tertiary. The low altitude ones are Middle to Late Tertiary age. Those profiles on the Deccan Traps, which were erupted in the Late Cretaceous, must have formed almost as soon as the Deccan basalt was erupted. The inversion of relief hypothesis accounts for older ferricretes being on plateaux and young ones being in valleys, and might account for the fact that high altitude ferricretes of the Deccan are older than low altitude ferricretes.

A very dubious 'lateritic profile' has been described from the Late Cretaceous/Early Tertiary of Japan (Iijima, 1972). This is a redbed of weathered pyroclastics, and it has been buried and folded so it is by no means related to the present land surface. It is mentioned here merely to provide further evidence of a weathering event around the Cretaceous/Tertiary boundary.

Calcrete

Carbonates are relatively mobile in the landscape and calcretes can apparently form more rapidly than ferricrete. Nevertheless, in some instances calcrete has given rise to inversion of relief in the same manner as ferricrete, and can be fitted into landscape evolution in the same way, like the example shown from Arabia in Figure 5.7. In Western Australia the Yeelirrie uranium-bearing calcrete is of Tertiary, probably Pliocene, age (Langford, 1974). In the United States there are also Tertiary calcretes with inverted relief (Miller, 1937).

In South Africa the Bushmanland plateau is characterized by calcrete (De Wit, 1988), which represents a period of climatic and tectonic stability. The calcrete affects the ultrabasic plugs of 70 million years old, and is cut by Miocene sediments, and so the calcrete formed sometime between 70 and 16 million years ago.

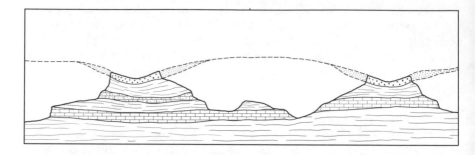

Figure 5.7 Inversion of relief in calcrete over limestone and shale bedrock in Arabia (after Miller, 1937).

Bauxite

Bauxite appears to be an end product of intense weathering when other components, including silica, have been leached away. It is thought to form under a humid tropical climate, and bauxite that is now in some other climate is usually related to some postulated former climate. Some bauxite is associated with limestone — karst bauxite — and some is formed on broad plains of apparently normal sediment or rock.

The bauxite of Weipa, Queensland, is at present in a monsoonal climate, but it is not clear why it is restricted to its present area when the same climate covers a much more extensive region on similar parent material. Under the bauxite are layers of ferruginous nodules, not usually cemented to form a ferricrete sheet, but their association with the bauxite is unclear. The age of the bauxite is not known but it is generally supposed to be Middle Tertiary. Some aspects of the bauxite, such as the distribution of the mineral gibbsite, which follows minor features of the present topography, suggesting a much younger age. This is in contrast to the usual idea that a long time as well as extreme weathering conditions are required to form a bauxite.

The isotopic composition of gibbsite from the Weipa bauxite indicates equilibrium with waters similar in isotopic composition to modern groundwaters of the region (Bird, Chivas and Andrew, 1989). Those authors assumed that the Weipa Bauxite was of Mid-Tertiary age, and concluded that a monsoon-type climate, similar to that of the present, has prevailed in the region since at least the Mid-Tertiary. Recent work (Pain, in press) suggests that the bauxite might be of only Quaternary age, and even forming at the present time, in which case the isotopic composition presents no problem.

The East Coast bauxites of India range in age from Late Cretaceous to Early Tertiary. According to Raman (1981):

the onset of monsoonal climate in upper Cretaceous promoted the process of residual chemical weathering. Chemical leaching was accentuated and optimised during upper Tertiary/Quaternary when there was copious rainfall with less of evapotranspiration, resulting in thick bauxite duricrusts.

The Antrim basalts of Northern Ireland erupted between 65 and 62 million years ago. They occurred in three main phases, and there are also minor breaks, marked by weathering, between flows. Some of the weathering intervals, especially in the lower series, appear to be long, and red beds, ferricretes and bauxites were formed. There are three types of deposits (Smith and McAlister, 1987):

1. Highly ferruginous red bauxites.
2. Siliceous grey bauxites.
3. Bauxites of mixed origin.

Bauxites have been recorded from many parts of the earth and of different ages. Some are as old as Devonian or even Precambrian, but these are not directly related to the present landscape. Buried bauxite of Jurassic age is associated with palaeokarst in Spain (Vera, 1988). Most bauxite that relates to present topography is probably of Tertiary age.

There are two main kinds of bauxite — lateritic bauxite (Bardossy and Aleva, 1990), and karst bauxites (Bardossy, 1989). About 10 per cent of the world's bauxite is directly connected with palaeokarst. According to Bardossy (1989) the role of karst can be summarized as follows:

1. Karst depressions serve as traps collecting bauxite from surrounding areas.
2. Karst depressions produce places of optimum drainage, promoting further desilicification of the bauxites.
3. Bauxites are protected in karst depressions from later erosion.

He stresses that the bauxite depressions were not fully developed when the first bauxite appeared. The evolution of the karst depressions goes hand in hand with the accumulation of bauxite material.

Sandplains of Western Australia

In parts of Western Australia the regolith consists of yellow structureless sand overlying rotted granite (saprolite) with an irregular contact (Figure 5.8). The first thought might be that in an arid environment wind-blown sands are to be expected, but the area is not truly arid, and the sands have none of the features of aeolian deposits. The distribution of the coversands corresponds to the distribution of granite bedrock, and the sands are a weathering product. Butt (1985) has described a profile through the sands and underlying saprolite. The sands result from extreme weathering in which even kaolinite is destroyed, but for some reason the usual bauxite minerals are not formed. The age of the coversand is not known, but this area has been land since the Precambrian so ample time is available for weathering, and rates of erosion are exceptionally low (Fairbridge and Finkl, 1978; Van der Graafe, 1981).

Figure 5.8 Sandplain profile in Western Australia exposed in a railway cutting. The upper mobile zone is highly gullied. Beneath it is an irregular surface, below which is massive saprolite with some corestones. (Photo: C.D. Ollier).

Weathering and derived sediments

Streams carry a bedload derived from their catchment, including pebbles released from bedrock by weathering and to some extent rounded by attrition. Around Canberra, for instance, modern streams may have pebbles of granite, schist, sandstone and quartz. However, when we look at ancient gravels in the same area, preserved on terraces, they contain only quartz. At the time the old gravels were laid down, the whole landscape was so deeply weathered that only quartz veins were providing pebble-sized material to the streams. This suggests that at that time deep weathering profiles covered the entire catchment. A similar story is found in the gold-bearing old river courses of Bendigo and Ballarat (the Deep Leads), which consist of a range of quartz-rich sedimentary types that are not produced by modern streams in the area. In contrast, near Armidale there are some Eocene old gravels that contain large pebbles of granite and even shale, but these are now reduced to clay by post-depositional weathering. Such deposits show, however, that the quartz-dominated sediments elsewhere are not a result of faulty observation: if other pebbles were ever present their remains can be recognized.

The conclusion from this is that in the Early Tertiary much of eastern Australia was so very deeply weathered that fresh rock seldom cropped out. Such extensive deep weathering suggests a period of low relief, little erosion, and a climate suitable for deep weathering — warm and wet. Early writers on Australian geomorphology noted the significant early phase in landscape evolution. Andrews (1910) postulated a Great Australian Peneplain, and Woolnough (1927) added the Great Australian Duricrust. In the absence of good dating, he presumed it to be of Miocene age, which is now seen to be too young.

What we know of the Mesozoic suggests that this would be an ideal time for extensive weathering and preservation of regolith. In fact, the Cretaceous appears to have been dominantly warm and wet all over the world, with no polar ice-caps and no deserts. Cretaceous, or possibly Mesozoic deep weathering could be a world-wide event, a unique event in the evolution of the world's landforms.

In the central uplands of Germany the development of the highest erosion surface began in the Mesozoic, and on the surface there are relics of deep weathering. Younger planation surfaces follow the main lines of the present-day drainage systems, and in some areas are covered in fluvial quartz gravels of probable Oligocene/Early Eocene age. As in Australia, these are thought to be derived from deep regolith (Andres, 1989).

Deep weathering and erosion surfaces

Erosion surfaces are treated at length in Chapter 7, including etchplains, erosion surfaces created in part by weathering. Some writers, such as Pavitch, Hack and Thomas are keen to see a steady state condition between weathering, erosion and planation surfaces: others see landscape-forming events occurring in succession.

Yet others see a former deep weathering profile, which is followed in landscape evolution by various degrees of stripping of the regolith to produce different kinds of etchplain, as shown in Figure 5.9. Again workers are divided on whether there is some sort of steady state in etchplain evolution, or if there is simply a major weathering followed by major erosion.

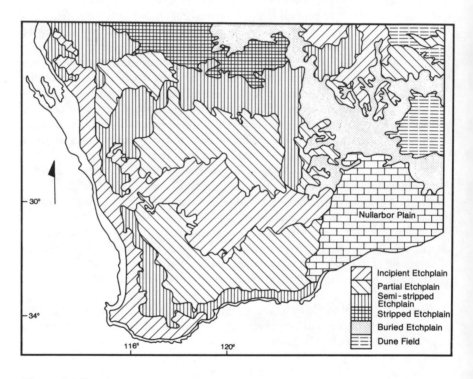

Figure 5.9 Etchplains of Western Australia. (After Finkl, 1979).

In Uganda a soil survey revealed several major units of regolith (Figure 5.10). In the south, deep weathering profiles are preserved in association with remains of an old erosion surface (the Buganda Surface), and the soils are formed on pre-weathered rock. To the north, remains of deep weathering are preserved, but there are no longer remnants of the Buganda Surface. In the north-west, the old deep weathering regolith has been stripped, and modern soils are forming directly on bedrock. The stripped area has been called the Acholi surface. A few deep troughs of weathering give occasional patches of pre-weathered regolith within the Acholi surface, and inselbergs are particularly common around the edge of the Acholi surface. These surfaces, and the regolith and soils upon them, certainly show historic landscape evolution and not a steady state. Absolute ages are not available for all the surfaces, and indeed they are diachronic to some extent. The Acholi surface, being related to the rift valley formation, is likely to have originated in the Miocene. The African surface is in places capped by Miocene volcanoes, but is likely to have originated in the Lower Tertiary at least, and the Gondwana surface is likely to be Mesozoic or older.

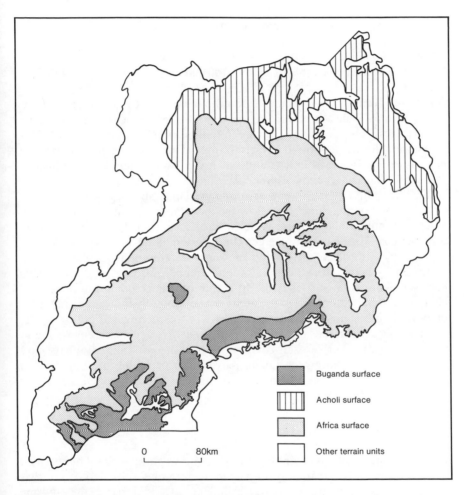

Figure 5.10 Stripping of regolith in Uganda. The Buganda surface has erosion surface remnants and deep weathering. The Africa surface is partially stripped. The Acholi surface is stripped of old regolith and modern soils are forming on essentially fresh rock rather than saprolite. (Source: C.D. Ollier, *Tectonics and Landforms*, Longman, 1980).

In northern Australia a similar story is found. A high erosion surface (the Ashburton surface) was cut across a wide range of bedrock. This was deeply weathered, and a silcrete was formed within the weathering profile. Later the area was eroded. The silcrete gave rise to a structurally controlled surface, but eventually the entire profile was stripped to produce the Tennant Creek surface (Hays, 1967; Wright, 1963).

Palaeosols

Some palaeosols remain at the ground surface, and are out of equilibrium with present-day conditions. For instance in Victoria, Australia, some weathered basalts have been at the surface for over 5 million years and have soils of a maturity not matched by those on flows 100 000 years old or those 10 000 years old.

Battiau-Queney (1984) reports that although deep weathering profiles are widespread in Wales, it is uncommon to find upper structureless horizons. Nevertheless, she discovered a true ferrallitic palaeosol on Carboniferous limestone. She also described a vertisol (tropical black soil) formed on impure limestone near Oswestry, England, which is now covered by periglacial colluvium (Battiau-Queney, 1987).

It is important to remember that although a landscape may be ancient, the soils can be young. Pillans (1991) described the situation on the Monaro Plateau of southern New South Wales where a great deal of work shows a long time-scale landscape evolution since the eruption of the 36 – 54 million-year-old basalts in the area. Prolonged erosional stability of the upland is also indicated by weak dissection of much of the landscape, and numerous enclosed depressions and lakes. Despite the abundant evidence of prolonged landscape stability, the soils of the area are unexpectedly thin and do not have properties consistent with prolonged Tertiary weathering and pedogenesis. Pillans proposes that the landscape is highly sensitive to climatic changes, and that much of the Tertiary soil cover on the basalt was stripped off by wind and water erosion during the Quaternary. Other palaeosols are buried, and though not part of present-day land surfaces, they do tell us about palaeogeomorphology.

Buried palaeosols

From the Silurian of Ohio, Kahle (1988) reported a gley palaeosol beneath an unconformity. It is remarkably similar to modern gley soils, indicating that conditions of soil formation have been similar from the Silurian to the present, despite the dominance of land plants in post-Silurian times.

Lehman (1990) described and compared palaeosols in Texas formed on outwash mudstones from above and below the Cretaceous/Tertiary transition which may mark an important environmental change. Early Paleocene soils are thick, black, and have deeper clay and thicker carbonate horizons than their Cretaceous counterparts. This suggests that the environment changed towards higher rainfall and cooler temperature. The change was abrupt, but lasted for a long time (about 2 million years) — not the sort of duration that might be associated with the currently fashionable asteroid collision.

Stable-isotopes and weathering

It may be possible to date regolith materials, at least in Australia, by stable-isotope composition, based on the assumption that

Systematic variations in the isotopic composition of meteoric waters, and therefore of the authigenic regolith minerals that formed in equilibrium with them, are a result of the continent's drift from high to low latitude and changes in global climate. (Bird and Chivas, 1989).

Ignoring a few exceptions, residual clays collected *in situ* from regolith profiles of post-Mid Tertiary age have $\delta^{18}O$ values between +17.5 and +21.3 per cent. These high values are consistent with Australia's northward drift to low (warm) latitudes during this time. Pre-Mid Tertiary clays have lower $\delta^{18}O$ values between +10.0 and +17.5 per cent. Values of $\delta^{18}O$ less than +14 or +15 per cent are thought to reflect weathering during much earlier geologic periods, perhaps Early to Mid-Mesozoic when Australia was at high latitude. Regolith profiles with low $\delta^{18}O$ clays (<+15 per cent) are widespread in Australia, and it is possible that a much greater part of the modern landscape than previously recognized developed in the Early or Mid-Mesozoic.

Stable isotopes may also be used to discriminate between weathering and hydrothermal alteration. Rising waters, steam and other emanations from deep in the earth move upwards through enclosing country rock, and bring about alteration, including argillation (the formation of clay). This is not weathering, but may look very much like it. Stable isotopes may discriminate between the two. On a conventional plot of oxygen isotopes, weathering products should fall on or near the so-called Kaolin Line, and hydrothermal alteration should fall in a different field. Alteration of the Bega Granite in New South Wales was attributed to hydrothermal alteration by Dixon and Young (1981), though Ollier (1983) thought the evidence favoured a weathering origin. Analyses of stable isotopes of Bega Granite clays fell exactly on the Kaolin Line, supporting the weathering hypothesis. Weathered granite is in places overlain by Paleocene basalt, so the weathering was presumably of Mesozoic age.

Yet the situation is not that easy. The kaolin deposits of south-west England were once thought to be of hydrothermal origin (e.g. Exley, 1958), despite having all the features associated with deep weathering. Sheppard (1977) showed from a study of oxygen and hydrogen isotopes that weathering, not hydrothermal alteration, produced the saprolites of Dartmoor. But Durrance et al. (1982) have now produced a hypothesis of clay formation, in which meteoric water is circulated as deep groundwater by geothermal heat, which could give weathering-type isotopic ratios although the process they envisage is essentially hydrothermal.

Chapter 6
Karst landscapes

Landforms dominated by solutional processes are called karst, after a limestone area in Yugoslavia. Limestone is by far the commonest rock which is soluble in normal conditions. Limestone develops a great many surficial landforms, which are far too young to be of interest in this book. There is also a tendency for drainage to go underground, forming caves.

Three kinds of caves can be envisaged. First, there are those that form by the action of water moving from the ground surface to the water table. These are called vadose caves. Second, there are caves formed beneath the water table. Here solution can take place in all directions — down, up and sideways — so caves have distinctive features. Cave passages can be circular in cross-section, as solution takes place equally in all directions, bell-holes can dissolve their way upwards into the roof of caves, blind passages can form (as there is no need for through-flowing water), and very irregular solutional hollows can be created at the rock surface. These are called phreatic caves and features. Third, there are caves formed by streams that flow, underground, along the water table. These can erode and dissolve their way down and sideways, but cannot erode or dissolve upwards except in rare floods. This action tends to make long caves that are roughly horizontal, even in steeply dipping limestone, and the caves have tributaries like normal surface streams. These caves are called epiphreatic. Once these indications are recognized, caves have diagnostic value in landscape studies, showing the position of former water tables.

In many areas the karst landscape appears to be fairly modern. This is true, for instance, in the caves of County Clare, Ireland, where most of the karst features can be related to post-glacial surficial landforms. Since they are formed in Carboniferous limestone, the mystery is, why are there no pre-glacial caves? Other caves do have a long-term history of evolution. In considering karst in long-term landscape evolution it is useful use to the term *palaeokarst*, which includes *buried palaeokarst*, ancient karst buried by younger sediments or rocks, and *relict palaeokarst*, present landscapes formed in the past (Jennings, 1971; James and Choquette, 1988).

Buried palaeokarst

The oldest known palaeokarst features are from the Transvaal, South Africa, where karst caves and cave deposits are 2 200 million years old, Early Proterozoic (Martini, 1981). Upper Proterozoic palaeokarst has been described from South Australia (Rowlands et al., 1980). Precambrian palaeokarst has been described from the North-West Territory, Canada (Kerrans and Donaldson, 1988). A great many karst features are present in the Middle Proterozoic rocks. The authors conclude that the relative abundance of grikes and the paucity of solution grooves (karren) may indicate either that the climate was temperate during karst formation, or that regolith covered the karst and was stripped away later. Ford (1989) described Proterozoic karst of Ontario and Quebec 1 400 million years old.

Middle Ordovician buried palaeokarst is reported from Quebec (Desrochers and James, 1988). Features include dolines (bowl-shaped depressions up to 3 m across), rundkarren (rounded solution channels) and kamenitzas (flat-bottomed solution basins).

Ordovician karst in the Appalachians include topographic highs, sinkholes and caves that extend to over 65 m below the unconformity. Near-vertical sinkholes are filled with carbonate breccia and gravels. The karst features have possibly controlled later geological features such as the emplacement of base metals and hydrocarbons (Mussman, Montanez and Read, 1988). Karstification occurred during early Middle Ordovician in Texas, and is buried beneath younger Palaeozoic rocks (Kerrans and Donaldson, 1988). Surface and subsurface karst features, including caves up to 15 m in diameter, have been reported from the Silurian of Ohio (Kahle, 1988).

Osborne and Branagan (1988) described many examples of palaeokarst in New South Wales, with karstification taking place in the Devonian, Permian, Early Tertiary and Late Tertiary–Recent times. An interesting biological association is that bats entered Australia in the Miocene, and became a significant source of phosphatic cave sediments.

Mississippian (Carboniferous) buried palaeokarst, including sinkholes and caves, is present in Wyoming and neighbouring states (Sando, 1988). The old karst topography has been reconstructed as an area of rounded hills with a relief of about 60 m drained by three main rivers. Late Mississippian karst is also present in Colorado, including caves with several cavern levels developed adjacent to major (100–200 m deep) valleys. Hundreds of lead–zinc–silver–barite deposits occur within karst solution features related to the Late Mississippian landscape (De Voto, 1988). In New Mexico, in contrast, Mississippian paleokarst is restricted to the top few metres of limestone and is little more than small intergranular cavities (Meyers, 1988).

In the Appalachians of Pennsylvania a lignite terrestrial deposit of Late Cretaceous age rests on over 60 m of what is presumed to be residue from underlying carbonate rocks (Pierce, 1965). The deposit is presumed by Pierce to have formed in a sinkhole, in which case there has been surface-lowering of hundreds of metres since the lignite accumulated at the surface in Late Cretaceous times.

Wright (1988) described two kinds of palaeokarst in the Carboniferous of South Wales. The first has densely piped and rubbly solution horizons, comparable to some present day humid karst. The other shows less well-developed karst, with associated calcrete, possibly indicating a semi-arid climate. Battiau-Queney (1986) described buried palaeokarst in South Wales that consisted of solution pockets, vertical channels and grooves formed sub-aerially, and then buried under loose sandy material of Neogene age derived from deep tropical soils.

Craig (1988) described Permian karst from Texas, where the unfilled caves, and the local distribution of karstic features indicates formation in freshwater lenses developed beneath a cluster of low-relief limestone islands. In southern Spain there are several palaeokarst terrains in the Jurassic, including caves 100 m below the related limestone top, lined with flowstone (Vera et al., 1988).

Relict karst

The Qattara Depression in Egypt is the largest and deepest of the undrained natural depressions in the Sahara. It is about 300 km by 150 km, and the lowest point is 134 m below sea-level. Albritton et al. (1990) believe it started as a river valley but its formation is mainly due to karstic processes in the Late Miocene. It may have been deepened later by deflation. The bedrock is Miocene limestone dipping gently towards the Mediterranean.

In the Jura of southern Germany there are karst solution fissures that reach depths of 12 to 15 m. They are filled with red or yellow loams, and sometimes contain fossils. The fissures were developed in the period from the Middle Eocene to the Pleistocene at about the same level and often close to each other, though each fissure contains fossils of only one age. This indicates that the surface in which the fissures developed has not been lowered since the formation of the oldest fissures. The fissures occur in that part of the Jura that was not covered by the Miocene sea (Bremer, 1989).

The Rhenish Slate Mountains include a karst area in Devonian limestone. There are some large karst pits which provide traps for sediments. The oldest sediments, dated by palynology, are of Lower Cretaceous age and fluvial origin (Schmidt, 1989).

In eastern Bavaria the Franconian Jura is a hilly upland of flat-lying Jurassic limestone, with a karst landscape. The karst developed in two stages (Pfeffer, 1989). After deposition of the limestone in the Jurassic the area became land in the Lower Cretaceous. Karst topography developed with steep karst towers and karst depressions up to 200 m deep. There was a polje (flat-bottomed karst depression) 5 km long and 1 km wide, and a network of underground watercourses was formed. In the Upper Cretaceous this landscape was buried under a cover of first fluvial and later marine sediments up to 250 m thick. A second phase of karst development began after uplift and continued from the Upper Cretaceous through the Tertiary and Quaternary to the present.

Evidence has begun to accumulate to suggest that some Australian karst landscapes are very much older than previously imagined. The Buchan area in eastern Victoria is a karst region that bears a lot of evidence about the post-Eocene history of the landscape (Webb et al., in press). Prior to basalt eruptions about 40 million years ago, the ancestral Buchan and Murrindal Rivers occupied broad valleys flowing north to south. The highest caves at Buchan formed as phreatic systems along vertical joints below the water table associated with this ancestral Murrindal River. Eocene basalt flows filled both valleys, and caused the development of twin lateral streams (see p.141), the Buchan and the Timbarra Rivers. These rivers cut down, lowering the water table and draining the pre-existing phreatic caves. A period of stillstand ensued, possibly in the Middle Miocene, with the development of a broad valley along the Buchan River. Extensive phreatic caves formed along joints in the limestone beneath the river and its tributaries. Another period of downcutting ensued, followed by another stillstand, which resulted in the development of river terraces along the Buchan and the formation of extensive horizontal epiphreatic caves, largely along the strike of the steeply dipping limestone. Uranium dates on stalagmites show that the caves were drained at least 86 000 years ago and perhaps over 350 000 years ago, by a final minor episode of downcutting.

Some of the Timor Caves in New South Wales have existed since before the Late Cretaceous, and much of the present landscape of the area had its origins prior to the extrusion of basalts about 73 million years ago (Osborne, 1986). Main Cave was formed by the excavation of both bedrock and palaeokarst sediment below the water table under phreatic conditions, but the palaeokarst sediments represent an extensive earlier cave filled with stalactites and stalagmites under vadose conditions. The creation of phreatic conditions after vadose conditions shows there was a rise in the water table, probably caused by blocking the neighbouring valleys with lava flows. A tabular basalt body in the Main Cave is interpreted by Osborne as a flow filling a cave at a time intermediate between deposition of the palaeokarst and excavation of the present cave (Figure 6.1). Osborne (pers. comm.) has evidence for a Late Cretaceous to Early Tertiary origin of the Bungonia Caves, and for Permian palaeokarst at Jenolan Caves, both in New South Wales.

The Chillagoe karst in Queensland is even older. The Chillagoe area lies about 180 km west of Cairns. It is characterized by towerkarst, with individual towers 100 m high, 4 km long and 1 km wide. The limestone is of Silurian/Devonian age and was deformed in Late Devonian/Early Carboniferous times. From the Mid-Carboniferous to Permian the area had ignimbrite eruptions (showing it was land) and granite intrusions. The rocks were then eroded and covered by Cretaceous sediments, and this formation covered karst towers, so the early karst features had already been formed at that time. The Chillagoe towerkarst had at least 60 m of relief 130 million years ago.

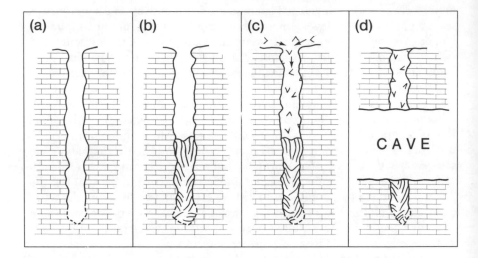

Figure 6.1 Possible mode of emplacement of basalt in Main Cave, Timor Caves, New South Wales.

1. Formation of narrow cave with connection to surface.
2. Deposition of secondary carbonates in the cave (stalagmites, stalactites, etc.).
3. Basalt enters the cave from a surface lava flow, and overlies the carbonate.
4. The present Main Cave develops between the dotted lines. (After Osborne, 1986.)

Erosion since then has added only 40 m to the local relief. By inference the towerkarst at Chillagoe, and by implication the caves, were well developed in the Early Cretaceous, and have probably been forming since the Carboniferous (Webb, 1988).

Ghizhou Province is situated on a karst plateau in subtropical South China. The karst developed in three stages and has been humid-tropical throughout that time (He, 1987). The stages are:

Dalou Mountain Stage. Denudation began in the Late Mesozoic, and in some of the early-formed basins sediments of Early Tertiary age are preserved.
Mountain Basin Stage. From the Miocene to Early Pliocene was a time of planation, creating the karst plateau with a few remnants of mountains and wide shallow valleys.
Wujiang Stage. Since the Middle Pleistocene there has been dramatic tectonic uplift, with rejuvenation of rivers and strong downcutting forming gorges.

Dragon Bone Hill in China has limestone caves including that in which Peking Man was found. The hill is in a syncline, and the development of caves is largely controlled by geological structure and lithology. The summit level of the hill is an erosion surface, on which there are Miocene beds containing fossils of the Hipparion fauna. Cave deposits just under the hilltop contain fish fossils of Late Miocene age, so there can be no doubt that the cave formation began at least as far back as Late Miocene and probably in Middle Miocene. These caves generally lack travertine or chemical deposits and it is possible that the caves were formed in the saturated zone before the old surface had been dissected by rivers. Later stages of cave formation in the Pliocene and Pleistocene are positively dated by mammalian faunas and magnetostratigraphic dating (Liu, Z., 1987).

In south-west China there is an area of perhaps the best developed towerkarst in the world. Its development goes back to the Cretaceous, and the evolution of the towers is influenced by the protection provided by Cretaceous red beds widely distributed both in the bottom of karst depressions, and on top of some of the towers (Yuan, 1987).

Salt karst

Karst is related to the geomorphology of soluble rocks, and is not strictly confined to limestone. Salt is another soluble rock, deposited in evaporation basins and attaining thicknesses of hundreds of metres. It also has the capacity to flow, making salt domes and related structures.

In the Texas Panhandle work on a nuclear-waste repository revealed a long history of salt geomorphology (Gustavson and Budnik, 1985). The salt was deposited in the Permian, and its distribution was affected by Permian faulting. Upper Permian salt was dissolved along the same line in later times, and is reflected in structural features and modern geomorphology. Salt solution has been a persistent feature from the Palaeozoic to the Neogene. A Middle Tertiary erosion surface has lows and closed depressions following dissolution along the same trend. Pliocene and Pleistocene lake deposits follow the same trend. Gustavson and Budnik wrote:

No single feature ... provides conclusive evidence of salt dissolution during the Neogene. However, there is a persistent pattern of structural and geomorphic features that can be best explained by dissolution of the Seven Rivers and Salado salts during the late Tertiary and perhaps as late as the Quaternary.

Chapter 7
Erosion surfaces

Many times in the history of the earth, folded rocks, dykes and intrusions of granite have been eroded to plains, known as planation surfaces or erosion surfaces. These are so obvious on geological cross-sections that their existence is hardly remarkable (Figure 7.1). Older ones are common in geological history as unconformities. Planation surfaces are equally obvious in the field, and are depicted on many photographs (Figures 7.2, 7.3, and see also Figures 8-12 a and b). 'Most people who are not blind or stupid can tell when they are in an area of relatively flat country: they can recognize a plain when they see one' (Ollier, 1981, p.152). For some reason erosion surfaces have come to be regarded as either mythical or old-fashioned concepts, especially by English geomorphologists, and even their existence is denied. We shall take it for granted here that they exist. What concerns us now is their age.

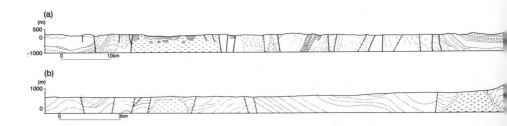

Figure 7.1 Geological sections to show planation surfaces cut across diverse geological structure. Despite some limited topographic expression, compared with the vast differences of structure in the bedrock the ground surface is essentially flat. The cross symbol indicates granite.
a. Dixon Range, Western Australia.
b. Canberra region, Australian Capital Territory.
There is no vertical exaggeration.

Figure 7.2. Planation surfaces in northern Uganda.
a. African Surface, Murchison Game Park.
b. Acholi Surface, near Aswa River.
Both surfaces are cut across steeply dipping metamorphic rocks.
(Photos: C.D. Ollier)

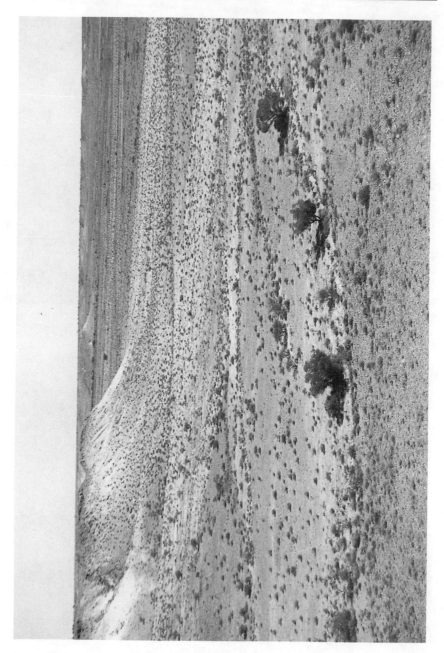

Figure 7.3 A planation surface near Coober Pedy, central Australia. The plain is cut across Cretaceous rocks, so is probably of Tertiary age. The silcreted plateau is a remnant of an older surface, also younger than the Cretaceous bedrock but older than the lower surface. The plateau surface may represent an old valley floor, part of an older planation surface. (Photo: C.D. Ollier)

Meaning of 'age'

In principle erosion surfaces can be dated within limits quite simply. Any erosion surface must be younger than the eroded rock it cuts across, and must be older than any rocks that cover it. In reality the dating of surfaces can be more difficult.

We must first consider what is meant by the age of an erosion surface. It takes time to form an erosion surface. Given some irregular initial surface, it will take time for the hills to be eroded, and perhaps more time still to reduce a subdued landscape to a flat one. Any hill will tend to be eroded to the level of the surrounding low country, that is to the local base level of erosion. A volcano on the plains of Uganda is in fact on a plateau at a height of over 1000 m, and will be eroded to the level of the plain eventually. Yet is the plain itself being eroded? On the flanks of the Rift Valley it clearly is, with valleys cutting down rapidly. The level of the Rift Valley lakes is here the base level of erosion. Ultimately all land can be eroded down to sea-level, which is therefore the ultimate base level. However, sea-level changes through time for many reasons, so is not an absolute datum limiting erosion.

Early ideas on the course of landscape evolution were expressed as the Davisian cycle of erosion. If an original plain were uplifted to form a plateau, rivers would first cut down towards a new base level, and for a while relief would increase and the landscape would become more rugged. When the streams had reached base level, downcutting would cease and the interfluves would be consumed as slopes became lower. Eventually a new plain (or nearly a plain, a peneplain) would be formed at the new base level (Figure 7.4a). Tectonic uplift could then start the process again. The landscape was conceived as a series of peneplains and partial peneplains. Many landscapes do not in reality look anything like this ideal, and we have no reason to suppose that the morphotectonic conditions of long period of erosion separated by abrupt periods of uplift are common.

(a)

Figure 7.4a The development of erosion surfaces:
According to Davis. A valley cuts down to a lower base level, and then the slopes become ever more gentle until a new erosion surface is formed.

Variations on this theme have been proposed. Penck believed that the interplay of tectonic uplift and erosion rates were reflected in the shape of valley sides, but this is now known to be an error: the shape of slopes depends mostly on the properties of the material and the nature of the slope-eroding processes. However, his proposal of parallel slope retreat has been shown to be correct in many instances, and was used as the basis of a landscape model by Lester King. King (1967) proposed that slope retreat created pediments, and when enough pediments were formed they constituted a pediplain (Figure 7.4b). This is akin to the peneplain of Davis, but with slope retreat rather than slope decline as the active process. Another suggestion by Crickmay (1933) was that plains called panplains are made by lateral planation by rivers.

It is very difficult to know how plains were originally created, but they can undoubtedly be seen in the landscape, so some authors prefer the term 'palaeoplain', which does not indicate the postulated process but merely

(b)

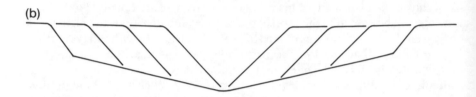

Figure 7.4b The development of erosion surfaces:
According to King. A valley cuts down to a new base level, and then the slopes retreat at a constant angle, producing a pediment at their foot. When a large area is formed by pedimentation it is called a pediplain.

means 'old plain'. Our concept of the age of an erosion surface depends to some extent on how we think it was formed. If we envisage some rugged landscape being worn down to a plain, it will become ever flatter with time, and we can do no more than suggest some arbitrary time when it became flat enough to be regarded as a planation surface.

If we think the planation surface is a pediplain, caused by parallel retreat of slopes, then the bits close to any residual mountains are younger than those further away. The surface was formed at different times (is diachronic) and if we put an age to the surface it must be the age of its initiation. Thus when a new sea-level was produced around Africa in the Early Tertiary, an Early Tertiary pediplain started to retreat inland from the coast. It may be still retreating into higher land in the continental interior, but near the coast may itself be under attack by the retreat of a younger pediplain.

If we have an old surface that has suffered much erosion, deposition, and perhaps volcanicity, but which is nevertheless distinctly flatter than the steep slopes which bound it, we only know it is older than its bounding escarpment. We can call it a palaeoplain without attributing any great precision to its age, and without being clear on all the processes that made it relatively flat. Even with these vague limitations, however, some writers have put ages to palaeoplains, generally meaning when they were first significant in landscape evolution. Hills (1975), for instance, suggests that the palaeoplain preserved in the Kosciusko Plateau in south-east Australia is Triassic to Jurassic in age. His reasoning is worth quoting at length:

The first appearance in south-eastern Australia of a land surface in any way resembling the present was in the Palaeozoic some 300 million years ago. By late Carboniferous and Permian, a continent composed of complex older formations had already been formed and strongly eroded before ice sheets advanced over the surface from the south-west during the Permian. Relics of the glacial advance in the form of tillites, outwash gravels, lake deposits and glacial pavements, now remaining in the western, central and northern parts of Victoria indicate that *the Permian landsurface was in places either the same as the present or close to it*, and elsewhere where deeper troughs existed these became largely filled with sediments.... The landsurface was, however, not

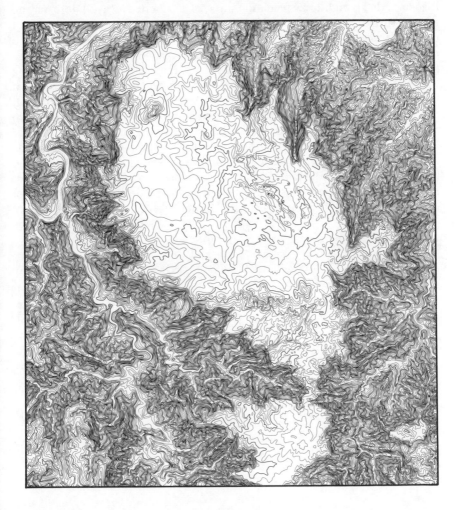

Figure 7.5 Contour map of the Carrai Tableland, northern New South Wales, showing an isolated high remnant of an erosion surface. Based on 1:100 000 topographical map, reduced.

yet an oldland, as there was considerable local relief, and the possible presence of an east–west belt of highlands is indicated by decreasing coarseness of the outwash gravels towards the north.... After the disappearance of the Permian ice sheets, an extensive and stable land area continued to exist throughout Triassic and Jurassic time, and a widespread palaeoplain was developed which may conveniently be identified as the Trias–Jura palaeoplain.

Dissected planation surfaces

Plateaux, that is high-level planation surfaces, are eroded away. There may be general surface lowering, reducing minor hills that rise above the plateau, but usually far more important is the erosion of steep slopes around the plateau edge. This is usually very obvious on contour maps, on the ground, and on air photographs and satellite images. The erosion can clearly be seen extending up valleys, and the norm is for valleys to cut down, widen and coalesce, so that much of the scarp consists of steep valley sides. The retreat of escarpments in this way may leave 'peninsulas' or small plateaux almost cut off from the main plateau, and eventually separate outlying plateaux (Figure 7.5).

In some instances most of the plateau may be removed, but remnants remain on very hard rocks such as quartzites. If these are steeply dipping, bevelled cuestas are developed. In highly eroded terrain these are the best indication of the former existence of an old erosion surface (Figure 7.6), for there is no way that the top of a cuesta can be flattened after downcutting of

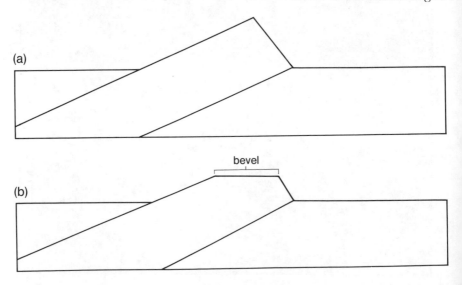

Figure 7.6 (a) A cuesta is a ridge formed by differential erosion of a dipping hard stratum. (b) A bevelled cuesta has a flat bevel that is a sure sign that an upper planation surface at about bevel level existed before differential erosion created the cuesta.

valleys around it. In some instances bevels can be followed around folded rocks (Figure 7.7).

Some examples of planation surfaces

In south-east England the Cretaceous Chalk is planated to give the flat plateaux of the Downs, which must be post-Cretaceous. A bevel cut in the Chiltern Hills, also Chalk, was dated by overlying sediments as Pliocene.

The Mendip Hills of Enland have a plateau top cut across steeply dipping Carboniferous Limestone, with a rather complex origin. According to T.D.

Figure 7.7 A bevelled cuesta, Carr Boyd Range, Western Australia. Note how the bevel can be traced around the nose of the anticline. (Photo: C.D. Ollier)

Ford (1989), 'The present upland surface of the Mendip Hills is thus a rejuvenated Triassic island surface, which has been subjected to slow phreatic solution, with marine planation in the Middle Jurassic, and some Pleistocene trimming.'

The Central Plateau of France consists of pre-Jurassic rocks which were folded, intruded and almost planed down and covered by transgressive Jurassic deposits. At the time of Alpine folding in the Tertiary the Central Plateau was uplifted and faulted. Uplift was greatest in the south and east sides facing the Alps. The faulting produced troughs in which Tertiary sediments are preserved, and volcanoes erupted in the Upper Tertiary and

Quaternary. Despite this lithological diversity, the surface of the Central Plateau is a monotonous erosion surface. It may consist of two parts (Sparks, 1972): a surface exhumed from beneath Jurassic cover, and a younger surface that started to form in the Eocene. This less perfect surface is also deeply weathered.

In the central German uplands peneplanation was largely complete by the Early Tertiary. In some areas such as the Eifel and the Ardennes, marine abrasion by the Cretaceous sea may also have played a part in the planation process (Andres, 1989). The south German scarplands lie between the Rivers Main and Danube, and include many denudation surfaces (Bremer, 1989). Uplift and denudation began in the lower Cretaceous, as shown in the Franconian Jura where the resistant limestones formed domal hills which were buried under Cretaceous marine sands, and must have existed before the Cretaceous transgression.

In the karst area of the Franconian Jura in eastern Bavaria the region had a marine inundation in the Upper Cretaceous. Between the Upper Cretaceous and Upper Miocene planation surfaces and low dome-shaped inselbergs developed at elevations close to base level, under conditions of deep weathering. The lowest erosion surface is at 480 m and follows main rivers. Three higher levels at 500 m, 540 m and 570 m occur independently of rock types and geological structures throughout the region (Pfeffer, 1989)

In the Central Plateau of the Iberian Peninsula the Hercynian massif is truncated by a very clear erosion surface, described by Molina et al. (1990) as 'the fundamental polygenic peneplain'. A weathering mantle formed on this surface was preserved by downfaulting in the Paleocene (58 million years ago), so the planation surface must be older and is probably Mesozoic.

In New Zealand the Central Otago Peneplain has been both warped and faulted (Stirling, 1990). This peneplain had a maximum slope of 500 m/10 km in the Miocene. As much as 1500 m vertical displacement of the peneplain is revealed by Tertiary sediments at varying altitudes. Thrust faults with a

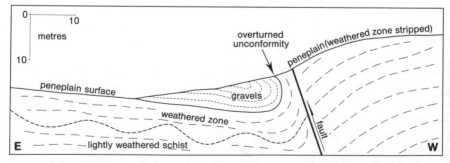

Figure 7.8 Deformation of the Miocene peneplain in Central Otago, showing asymmetrical fold, associated reverse fault displacement, preservation of quartz gravels and deep weathering on the downthrow side, and stripping of the weathered zone on the upthrown side. (After Stirling, 1990.)

displacement of tens of metres have warped and even overturned younger gravels, and led to stripping of weathering profiles on the upthrown block (Figure 7.8).

The southern edge of the Colorado Plateau, often called the Mogollon Rim, marks the southern termination of Permian cliff-making strata. According to Peirce et al. (1979) a plateau edge escarpment had evolved by Mid-Miocene times, prior to the onset of Basin and Range faulting and the development of the modern Grand Canyon. The preservation of Cretaceous marine strata along portions of the present plateau physiographic edge at elevations over 2 km indicates uplift relative to sea-level, but not the time of the uplift.

Bradley (in Madole et al. 1987) has provided an extensive review of the erosion surfaces of the Colorado Front Range. Most people now believe that a single, widespread erosion surface exists in the Colorado Front Range, commonly named Rocky Mountain, Sherwood or Late Eocene Surface. Bradley believes this consensus has regional significance far beyond the Front Range, that it may have had much greater regional extent, and it may transgress time (in which case Late Eocene is an unsuitable name). Questions remain about how the surface was formed, how much it has been lowered in post-Eocene time, and the significance of small surfaces that have been called the Flat-top Surface.

Graf et al. (1987) reviewed the geomorphology of the Colorado Plateau. The sequence of landscape-forming events is:

1. The uplift of the plateau margin during the Laramide (Late Cretaceous) earth movement, creating a broad erosion surface 'much like the surface presently observed over much of northern Arizona'. 2. The Laramide Surface was buried by up to 90 m of Rim gravel. 3. The Laramide was followed by Late Eocene–Early Oligocene tectonic quiescence and regional weathering. 4. The Colorado River began to emerge as an integrated drainage system in Late Miocene to Pliocene times.

In brief, the broad geomorphic framework of the Colorado Plateau appears to have been shaped by Laramide cliff recession under climatic and base level controls very different from those associated with Pliocene canyon incision.

In Wyoming Evanoff (1990) showed that the Eocene surface was quite rugged, and valleys were filled in with the Oligocene White River Formation. This formation is little deformed, ruling out post-Oligocene uplift as a source of modern relief. The regionally widespread low-relief summit surface was developed in the Miocene, and is not part of the late Eocene unconformity.

The complex and numerous erosion surfaces of the Sierra Nevada, United States, are discussed by Bovis (1987). It seems that the Mountain Canyon stage was well under way by 3 million years ago, so the earlier Mountain Valley and Broad Valley stages must go back to the Pliocene and Miocene respectively. It is notable that Bovis stresses the significance of the study of these erosion surfaces to geomorphological theory and thought, especially in relation to long-term landscape evolution on a time-scale of millions to tens of millions of years.

This is due in large measure to the availability of both erosional and depositional evidence, including datable volcanics. There are few other areas of North America in which geomorphology and regional tectonics have been so closely meshed, and in which denudation chronology has been able to fulfil a role beyond landform description.

According to Evanoff (1990) the age of the Rocky Mountain sub-summit surface varies from place to place. In central Colorado it is demonstrably late Eocene, but in much of the Laramide and Medicine Bow mountains it is Miocene.

Even in the Coast Mountains of British Columbia the present high relief is a product of rapid Late Cainozoic uplift. 'A mature erosion surface of moderate relief existed over most of southern British Columbia, including the southern Coast Mountains, in early to middle Miocene time' (Clague and Mathews, in Muhs et al., 1987). The Miocene landscape was characterized by broad low valleys separated by gentle slopes rising to narrow divides and summits 300 to 500 m above the low ground. Remnants of lavas erupted on to this surface when it was near sea-level are found today on a few summits up to 2.5 km elevation. Additional support for rapid recent uplift comes from an analysis of the present physiography of the Coast Mountains. Over a large area in the south, summits form a broad plateau-like envelope at 2 to 3 km elevation that may represent the Late Miocene erosion surface (Parrish, 1983). To the north the summit surface is lower and more irregular, suggesting greater antiquity of relief.

The Gran Sabana on the Guayana Shield in south-eastern Venezuela is mainly developed on gently folded Precambrian quartzites and conglomerates. Two planation surfaces are developed: the Auyuan-tepui (2 000-2 900 m), and the Wokien (900-1 200 m). Briceno and Schubert (1990) discuss them as follows:

assuming that Mesozoic climate was generally warm and dry, providing long periods of stable climate, that the Guayana Shield was situated in the tropics since the breakup of Pangaea, and that the Wonken and Auyuan-tepui Surfaces are the best-formed of all the planation surfaces, can be taken as tenuous evidence for a Mesozoic age of initiation for these surfaces.

In Wales planation surfaces have been recognized at a variety of heights, some of which are very clear. Of more significance than these partial surfaces is the the upland plateau surface of Wales, the Welsh 'tableland'. E.H. Brown (1960) recognized three main planation surfaces, all of subaerial origin, of which the 'High Plateau' is the tableland proper. He considered it dates only from Miocene times. In contrast, Jones (1931) postulated that the Welsh surface may represent a Triassic erosion surface which may have been covered for a long period by Mesozoic strata, so the High Plateau is essentially exhumed by Tertiary erosion. Battiau-Queney (1984) pushed the age back farther when she wrote. 'I believe that *the geomorphological evolution leading to the present landscape started much earlier, in Devonian times'.* (her italics). She believes also that Lower and Mid-Jurassic seas submerged the major part of Wales, and that the Welsh massif emerged between upper Jurassic and Cenomanian — a period important in the development of the future Atlantic margin.

In the east Victorian Highlands (Australia) there is direct evidence of the age of the Trias–Jura palaeoplain of Hills (1975) in the Benambra region. Trachytic lavas and tuffs, with associated intrusions, are dated at a little over 200 million years, which places them within the Triassic.

The preservation of flows and tuffs indicates that the surface on which they rest is also Triassic, and although considerable erosion of the large Triassic volcanic cones has since taken place, the gently rolling, downlike topography of the upland surface around Benambra may be regarded as a somewhat modified relic of the Triassic palaeosurface.

For comparison, here is Lester King's description of the Gondwana planation surface of Jurassic age:

In the cratonic areas of South and East Africa, southern India, Brazil and elsewhere small relicts of this planation still survive upon the highest and most ancient watersheds. There, high above the well-planed 'Moorland' early Cenozoic planation (III), they have been continuously exposed to the winds and the weather since they first took shape in Gondwana at least 170 million years ago. Naturally, such relicts are very rare in existing landscapes: yet despite their great age, *they have not been buried and resurrected at any time in their history.* (King's italics)

In the Davenport Ranges of northern Australia, plateau remnants are preserved on steeply dipping quartzites. In the valleys between, river terraces are capped by gravels, which are of Cambrian age (see p.22). The Davenport planation surface is therefore likely to be Precambrian, possibly correlating with the Kimberley planation surface. Erosion surfaces further south, such as those on the Macdonnell Range, were once thought to be Cretaceous (a daringly old age at the time) but they may well be older. The planation surface of the Yilgarn area of Western Australia is of unknown age, but the deep weathering which is older than the Eocene valleys shows it is very old. It may be Precambrian, and has certainly been a land area since that time.

The Appalachian Plateau of the United States has long been a controversial area for the study of erosion surfaces. The hard bands of the folded rocks are planated, and the major drainage is superimposed (p.33). There are many arguments about the number of erosion surfaces, their origin and their age. The oldest estimate is Permian, by Meyerhoff and Olmsted (1963).

According to Rodgers (1983), sinkhole deposits with fossils of Paleocene and Late Cretaceous age are found on surfaces hundreds of metres below the crestal ridges of the Appalachians. The ridges must be considerably older. The sinkhole hypothesis requires solution of hundreds of metres of limestone beneath the fossiliferous lignite beds. Pierce (1965) describing such a deposit in Pennsylvania also considers the possibility that it is a simple surficial deposit and not a sinkhole deposit: 'If this be true, the deposit accumulated in the valley floor of a "partial peneplain" and the major features of the present topography are relics of Late Cretaceous time.'

Some African erosion surfaces may be related to rift valleys and drainage pattern evolution (p.40). The oldest surfaces are at least Mesozoic. Other small fragments of surfaces are being exhumed from beneath Miocene volcanoes such as the Mbale Surface emerging from under Mount Elgon in Uganda. Such erosion surfaces are also warped, and the post-erosion attitude can be mapped and depicted in sections (Figure 7.9).

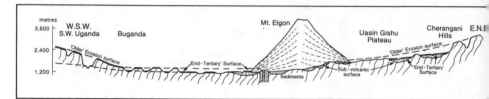

Figure 7.9 Erosion surfaces warped by earth movements related to the formation of rift valleys in East Africa, also showing the relationship to the Miocene Mount Elgon. (After King, Le Bas and Sutherland, 1972.)

The Kimberley Plateau of north-west Australia is largely preserved on sandstones of low dip, but in some areas tightly folded rocks give rise to bevelled cuestas. In places these have been glaciated by a Precambrian glaciation, so the plateau was already in existence in Precambrian times and has been land ever since (Ollier, Gaunt and Jurkowski, 1988).

Many of the erosion surfaces discussed so far are remnants preserved on high ground. In contrast some erosion surfaces are made more obvious by large hills or inselbergs rising above them. A classic example is Ayers Rock in central Australia (Figure 7.10). Paleocene lake sediments lie on the plain between Ayers Rock and the even larger mass of the Olgas (Twidale and Harris, 1977), so Ayers Rock looked very much as it does today in Paleocene times, 60 million years ago. The erosion surface itself is likely to have a very much greater age. About 250 km to the north near Hermannsberg unconsolidated alluvium has been found to contain Permian pollen (E. Truswell, pers. comm.), so the plain could be as old as that in its main features and even in some of its details.

In West Nile Province, Uganda, careful geomorphic work by Hepworth (1964) showed that what in some areas appear to be two surfaces are in fact one surface, downwarped along a monocline. Erosion on the steep monocline produced an escarpment and a line of inselbergs, but the upper and lower surfaces are really the same. The age of the warping is not known but is likely to be Early Tertiary, like the associated rift valley faults.

Near Sydney, Australia, the Lapstone Monocline separates an upper and lower surface of the same age, just like the West Nile monocline described

Figure 7.10 Ayers Rock, central Australia. Paleocene sediments are found on the lower erosion surface, so that even in Early Tertiary times Ayers Rock appeared very much as it does today, and the planation surface was already present. (Photo: Australian Information Service)

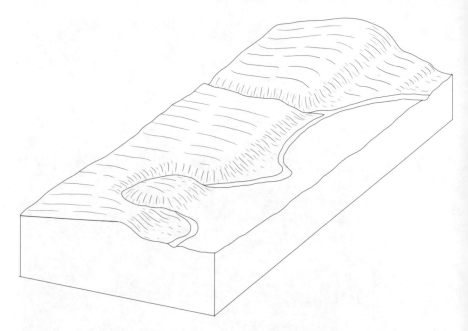

Figure 7.11 The Lapstone Monocline. A sketch showing the Nepean River flowing in an antecedent course across the monocline, showing that the river had that course before earth movement created the monocline. The monocline is of Early Tertiary origin, so the river, and the plain it flows upon, which was warped to form the monocline, are even older. (Source: C.D. Ollier)

above (Figure 7.11). Palaeomagnetic work shows that the monocline probably exceeds 15 million years old (Bishop, Hunt and Schmidt, 1982), but more detailed work shows the structure is complex in detail (Branagan and Pedram, 1990), with a fault system rather than a simple monocline. Most of the deformation took place in the Early Tertiary. The superimposed Nepean River winds across the monocline, and has maintained its course through this long history. Of course the plain on which the Nepean was flowing before it was warped into a monocline must be even older.

Etchplains

An etchplain is an erosion surface formed by first deep weathering, and then erosion of the weathered material to form a lower surface. Different types of etchplains have been defined (see Thomas, 1989; Ollier, 1984) and they have been mapped in different ways. Nevertheless, it seems to me that the age of an etchplain must be younger than the age of the preceding weathering, which must be younger than the age of the preceding planation of unweathered rock. However, Thomas (1989) believes that weathering and erosion may be in dynamic equilibrium, on the model proposed by Hack (1960). This idea is discussed again in Chapter 14.

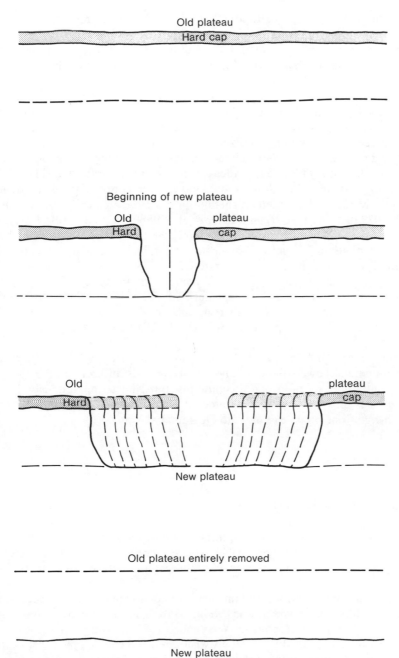

Figure 7.12 The evolution of the Western Australian plateaux according to Jutson (1934). The original meaning is not entirely clear, but probably Jutson implied a phase of deep weathering to a horizontal weathering front, followed by stripping to produce the New Plateau. It is also interesting that Jutson anticipated parallel retreat of slopes in his diagram. (After Jutson, 1934).

The ferricreted deep weathering profile of the Gulfs region in South Australia developed during the Mesozoic. Its stripping to expose the high etchplain that is such a prominent feature of the Mount Lofty Range, east of Adelaide, did not begin until the Ranges had been uplifted, beginning in the latest Cretaceous or Paleocene (Twidale, 1990).

Deep weathering is also present in Western Australia. An early account by Jutson (1934), first published in 1914, is illustrated diagrammatically in Figure 7.12. Beneath the Old Plateau — an ancient erosion surface — there is possible deep weathering to a horizontal weathering front. Later erosion cuts down to a new level, and then valley-widening leads to the formation of the Younger Plateau. These early ideas seem to incorporate the ideas of weathering and erosion (etchplanation), and also the notion of parallel retreat of slopes. Modern ideas of etchplanation for the same area are presented by Finkl, 1979 (see Figure 5.9) and Ollier et al. (1988b), who describe Eocene valley fill on the New Plateau, and Mabbutt (1988), who described it from the Wiluna-Meekatharra area. He describes the Old Plateau as in part an exhumed pre-Permian surface of low relief, of Gondwana age. Landforms above the Old Plateau have maintained their relief during this circumdenudation and there is no regional evidence of their isolation by major escarpment retreat. The New Plateau surface extended by stripping of saprolite and is an etchplain (as is the Old Plateau under the genesis postulated).

Kroonenberg and Melitz (1983) described the stepped landscape of Surinam, which they attribute to lithological contrasts differentiated by deep weathering under humid tropical conditions, but later stripped of saprolite. Old surfaces only survive where protected by a duricrust. Erosion surfaces range in age from Jurassic or older to Quaternary.

Erosion surfaces and duricrusts

Many plateau remnants of erosion surfaces are capped by duricrust (Figure 7.13). A question is: was the original high plateau entirely covered by duricrust before dissection, or did patches of duricrust protect selected areas for preservation?

Several authors have discussed the close connection of erosion surfaces and duricrusts, like Kroonenberg and Melitz (1983) as mentioned above. An extreme case was the account by Woolnough (1927), who proposed that a Great Australian Peneplain was deeply weathered and covered by duricrust in the Miocene. Certainly a caprock of duricrust must help to preserve remnants of erosion surfaces, but present-day opinion suggests that the duricrust was never present over the entire former erosion surface. Some (e.g. Milnes et al., 1985; and Ollier and Galloway, 1990) suggest that ferricretes form on valley lowlands and come to be on plateaux by inversion of relief. Duricrust outcrops

Figure 7.13 Duricrusted plateau, with silcrete and minor ferricrete on Cretaceous marine sediments, Coober Pedy, central Australia. (Photo: C.D. Ollier).

around plateau remnants are very eye-catching, but there is no reason to suppose that they were very much more extensive in the past than they are today, and there is no justification for casting imaginary surfaces to link all the plateau remnants and assume that duricrust covered it all. Improvements in dating methods will improve our knowledge of the relationship between duricrusts and erosion surfaces. Even now we know that many duricrusts are of Early Tertiary and Mesozoic age, which of course tells us something of the age of the erosion surfaces on which they are found.

Multiple planation surfaces

Some authors have described multiple erosion surfaces, including some examples mentioned earlier.

Engeln (1963) recognized several erosion surfaces in the Dolomites of Italy. After initial uplift in the Lower Tertiary an erosion surface was formed in Oligo-Miocene time, which is extensively preserved on limestone and dolomite. Renewed uplift in the Miocene and Lower Pliocene led to downcutting of about 1000 m, followed by formation of a new planation surface in the Upper Pliocene. This surface has since been uplifted and ranges in altitude from 1 900 m to 2 300 m.

Table 7.1 Global planation cycles and their recognition

Planation cycles

	OLD NAME	NEW NAME	RECOGNITION
I	Gondwana	The Gondwana planation	Of Jurassic age, only rarely preserved
II	Post-Gondwana	The 'Kretacic' planation	Early mid-Cretaceous age
III	African	The Moorland planation	Current from Late Cretaceous till the mid-Cainozoic. Planed uplands, treeless and with poor soil.
IV		The Rolling landsurface	Mostly of Miocene age, forms undulating country above younger incised valleys
V	Post-African	The Widespread landscape	The most widespread global cycle, but more often in basins, lowlands and coastal plains than uplifted by recent tectonics to form mountain tops. Pliocene in age
VI	Congo	The Youngest cycle	Quaternary in age represented by the deep valleys and gorges of the main rivers

Lester King (1983) described a series of planation surfaces (pediplains) in Africa, and then claimed to recognize the same surfaces around the world, indicating a common earth history in the different continents. His surfaces and the ages assigned to them are shown in Table 7.1. In earlier work he used different names for some of the surfaces, and since other workers often use the old names I have shown the rough equivalents in Table 7.1, but the new scheme has six surfaces compared with the five of the old system. I have had the pleasure of looking at these surfaces in the field with Lester King in both Africa and Australia, and we agreed to differ. I see only one palaeoplain (with a few residual hills) and a younger surface related to the present continental edge. He could see a flight of surfaces.

A detailed review of the study of erosion surfaces in southern Africa was provided by Partridge and Maud (1988). They show a wide range of opinion on the number and age of surfaces. They themselves doubt the existence of the oldest surface (Gondwana, or Jurassic) and consider the African surface to be the dominant feature in the landscape. 'We cannot find any conclusive evidence for the preservation of any subaerial erosion surface dating from before the breakup of Gondwanaland.' Instead they follow Hawthorne (1975), who presented evidence (based on the study of volcanic pipes) for removal of 300 m of material from the summits of the Lesotho Highlands since the Late Cretaceous. Nor do they find evidence of the post-Gondwana surface, though they point out that gaps in the marine sediments offshore in the Early Cretaceous and Mid-Cretaceous are correlative of such surfaces. They believe the African cycle was initiated by the breakup of Gondwanaland and persisted until its termination by marginal uplift in the Early Miocene. They believe that Africa had a high position before breakup, so that after breakup there were two base levels in the interior and marginal parts of the continent, so erosion surfaces of the same initial age but different elevation have been planed on opposite sides of the Great Escarpment. They say the African cycle was terminated (on the eastern side of the continent) by uplift of 200 to 300 m in the Early Miocene initiating the post-African surface, and in the Late Pliocene there was uplift of 800 to 1000 m associated with marginal tilting. The south and west continental margins were uplifted by smaller amounts.

In the southern coast region of South Africa there are at least three well-developed elevated planation surfaces fringing the coastal mountain ranges (Jacobs and Thwaites, 1988). All have been formed by sub-aerial processes (based on presence of fluvial and aeolian sediments, silcretes and the lack of marine sediments). Lignites are probably Miocene, so the oldest surface is pre-Miocene and possibly as old as Mid-Cretaceous. This of course makes the backing mountains much older. The main coastal platform is Mio-Pliocene, with upheaval in the Pliocene.

The north-western part of Cape Province appears rather different (De Wit, 1988). Exposed volcanic plugs, and fission track dating of basement rocks suggests an early aggressive erosional phase from the Late Jurassic to Early Cretaceous followed by a long interval of stability and production of mature pedogenic profiles.

Similar multiple erosion surfaces have been recognized in many other parts of the world. In India, for example, Rai (1987) described essentially the King surfaces from the Meghalaya Plateau, in the northern zone of the Deccan facing the Himalayas. The surfaces are as follows:

1. Gondwana surface   1500–1800 m
2. Cretaceous surface   1200–1500 m
3. Eocene surface   900–1200 m
4. Pleistocene surface   600– 900 m
5. Pleistocene to Recent surface   300– 600 m

In northern Portugal the granitic Serra da Peneda are part of the marginal swell of north-west Iberia. Five planation surfaces were recognized by Coude-Gaussen (1981) and dating was attempted by correlating products of sub-aerial denudation and the offshore deposition near the western continental margin of Iberia. The surfaces are:

I.   Lower to Middle Jurassic
II.   upper Jurassic to Middle Cretaceous
III.   Upper Cretaceous to Middle Eocene
IV   Upper Eocene to Upper Oligocene
V.   Miocene to Pliocene

Levels I and III are the most extensive. She proposes a model which associates the epeirogenic processes of the continental swell with the subsidence of the passive Atlantic margin. The discontinuous uplift of the mountains seems to be related to the maximum spreading phases of the North Atlantic (see Chapter 8, p.104).

In the northern part of the Central Cordillera of the Colombian Andes extensive remains of three planation surfaces have been recognized (Page and James, 1981):

The pre-Cordillera Central Surface, over 3 000 m
The Cordillera Central Surface, 2 500–3 000 m
Rio Negro Surface, around 2 200 m

The oldest surface has an age of 22–18 million years. The Rio Negro surface was eroded about 4 million years ago. The western Cordillera of Colombia, unlike the other cordillera in Colombia, consists of oceanic crust that was accreted to the South American Plate in the Upper Cretaceous, but even here uplifted erosion surfaces are extensive (Padilla, 1981).

Mortimer (1973) recognized four episodes of landscape evolution in the Atacama Desert of Chile, each of which led to formation of a planation surface. The first began with the elevation of northern Chile in the Late Mesozoic. Each subsequent phase started with incision of drainage into the preceding landscape. Volcanic rocks provide dates. The present phase of canyon-cutting commenced in the Upper Miocene.

Melhorn and Edgar (1975) argued for four world-wide erosion surfaces, and produced a theoretical geodyamic hypothesis for their formation. They believed there were four periods of continental quiescence in the Mesozoic and Tertiary when the surfaces originated, each lasting about 25 million years.

Problem surfaces

Early enthusiasm for erosion surfaces probably caused many phantom surfaces to be reported in the literature. In New South Wales perhaps the first reported peneplain was that of the Mole Plateau. This is indeed a splendid plateau bounded by steeper slopes and is cut across granite. However, as explained on p.156 (Chapter 11) the Mole Granite is a flat-topped, mushroom-shaped intrusion that was emplaced beneath a cover of only half a kilometre of cover rock that was 'floated off', and the top of the granite might have been above the general ground level even at the time of intrusion. It is thus a structural surface, and its level was never related to a base level as is a true planation surface. Other peneplains identified in the same region, such as the Bolivia Surface, are also granite tops rather than planation surfaces related to some base level.

At the extreme, the search for erosion surfaces can lead to conjectural surfaces cast across rugged landscapes. In a genuinely rugged landscape there is no point in inventing surfaces to pass over the top of the landscape, as even though it is clear that former higher land existed, there is nothing to show it had a flat surface. There are some places, however, where although distinct plateau remnants are lacking, the accordance of summits, and especially long ridges is too good to result from chance erosion of dipping rocks. Figure 7.14 shows how downcutting of valleys can lead to lowering of divides, which might be connected to reveal hypothetical or mythical erosion surfaces. Yet this two-dimensional diagram of a sort of dynamic equilibrium does not display the full evidence available for judgment. In the third dimension ridges would be free to vary along their length, as is sometimes seen in dissected folds. Accordant levels along the strike are reasonably good evidence for the former existence of an erosion surface. The Appalachians provide good examples in my opinion, but many workers in the area refuse to accept them. How far the interpretation can be carried is a matter of scientific taste.

One of the possibly dubious erosion surfaces is the Gipfelflur, a surface cast across the European Alps. Rutten (1969) described it thus:

From afar one always sees how the individual tumble of jagged peaks, so impressively irregular at close sight, approaches a strikingly even upper limit. This summit level passes over the tops, which are in most cases residual mountains of some former erosion surface. Many individual peaks are eroded below the surface, and a few stick above it, and there are wide undulations of the summit plain itself. But as an overall feature this summit plain is a very real thing indeed.

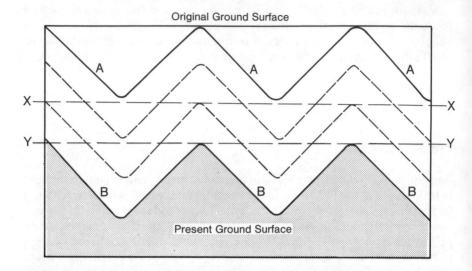

Figure 7.14 Cross sectional view of a possible sequence of landscape evolution, with relatively closely spaced streams relative to the amount of downcutting. It is assumed that there was an original flat planation surface. When erosion has reached the stage of AAA, the old surface could be reconstructed by drawing a surface across the interfluves. At later stages this technique would not work and "surfaces" XX and YY never did exist as planation surfaces. This two-dimensional approach, however, does not always limit the technique, and in the real world the levelness of interfluves can add another dimension and make the interpretation of old erosion surfaces much more plausible, as in Figure 7.9.

In the case of the Alps the Gipfelflur was upwarped during Late Pliocene and Pleistocene times. I think I have seen a Gipfelflur in the Owen Stanley Mountains of New Guinea, in the New Zealand Alps (with Mount Cook rising above it), and in the Himalayas (with Mount Everest rising above it), but whether in any of these locations I was seeing a real geomorphic entity or a trick of perspective in not clear.

Augustinus (in press) studed the Fiordland area of New Zealand, an area of intense glacial erosion. Nevertheless, he was able to produce a map showing a summit level envelope (Figure 7.15) which is a kind of planation surface or Gipfelflur. Its age is probably between 0.5 and 2 million years. The Gipfelflur is a useful concept, but perhaps not amenable to proof.

Exhumed erosion surfaces

The geological column is full of unconformities — surfaces separating older, often folded, rocks from overlying younger, often near horizontal rocks. Some unconformities are planar. They may have been formed by coastal erosion during a marine transgression, or by terrestrial erosion to a peneplain. The overlying rock may be marine (commonly a conglomerate if marking a transgression) or terrestrial.

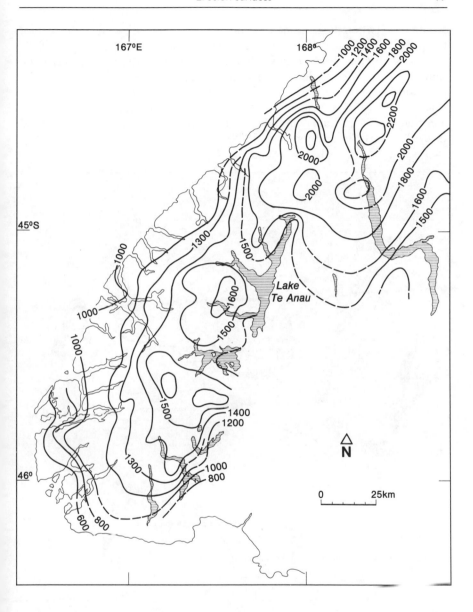

Figure 7.15 Summit surface envelope, or Gipfelflur map, of Fiordland, New Zealand, (After Augustinus, in press.)

If the overlying softer rocks are removed by erosion to reveal the old unconformity, an exhumed erosion surface is produced. How this relates to total landscape evolution is sometimes debatable, but depends on whether a thin cover has been stripped (when the old erosion surface plays a large part in modern topography), or whether hundreds or thousands of metres have been stripped off (when the exhumed erosion surface is almost an accidental part of the present landscape, like any other structural surface).

The Kimberley Surface described earlier is one where little has been removed, and the modern surface corresponds in some detail with the Precambrian surface. The Cambrian Peneplain in southern Sweden had a thicker cover, but corresponds to the present land surface over such a large area that it has a far more than coincidental effect on the present landscape and its evolution (Lidmar-Bergstrom, 1988).

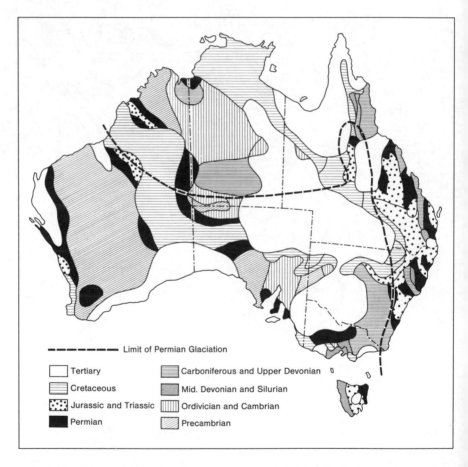

Figure 7.16 Maximum age of exposure at the groundsurface of various regions of Australia. (After Beckmann, 1983)

The landscape of the Canadian Shield is the product of a combination of Precambrian peneplanation and Pleistocene Glaciation (Shilts et al., 1987). The overall surface of the Shield is an exhumed pre-Palaeozoic erosion surface (Ambrose, 1964) that slopes gently from elevations of 400–600 m near its margins to a central depression occupied by Hudson Bay. The total amount of glacial erosion is likely to be less than 20 m, so the present bedrock morphology of the Shield is not substantially different from that which existed before glaciation.

An exhumed Sub-Triassic planation surface occurs in south-west Liverpool Land, Greenland (Peulvast, 1988). It is preserved on a 10 km-wide westward sloping plateau of Liverpool Land. Tors and red coloured saprolite on weathered granite indicate exhumed weathering front and partial preservation of Triassic saprolite (Coe, 1975). This surface may be an exhumed etchplain.

Exposure of land surfaces, drainage age and erosion rates

Although not strictly planation surfaces, land surfaces of a given age of exposure can be considered here. For Australia Beckmann (1983) has produced a map showing the maximum age of exposure of regions, based on palaeogeographic maps (Figure 7.16). It is evident that large areas date back even to the Precambrian. A further major division is the limit of Permian glaciation, for all pre-Permian surfaces south of the line may be partly stripped, and then re-exposed to weathering and erosion.

Elsewhere maps of this surface age are not available, but there are individual statements about the age of exposed surfaces. For instance, the Wright Valley in Antarctica is one of the largest ice-free areas on the continent. Within it are high altitude extremely fretted and weathered land surfaces that have been ice-free possibly since the Early Miocene (Campbell and Claridge, 1988).

A related kind of map has been prepared by G. Wilford (pers. comm.) which shows the age of inception of drainage in Australia (Figure 7.17). When the sea and ice sheets have retreated for the last time a drainage system will be initiated on the newly exposed land, and this may persist in some form right down to the present. In allocating ages Wilford has mainly ignored relatively local drainage changes arising from warping, faulting, river capture, volcanic, glacial and aeolian activity. The map shows the very great age of much of Australia's drainage, and Wilford has provided the following commentary.

1. Proterozoic drainage

The Kimberley area has been land throughout the Phanerozoic and was apparently little affected by Permian glaciation. Drainage radiates from the area with the highest accordant summits suggesting it was initiated on a north-west-trending dome-shaped uplift which pre-dates the Late Proterozoic glaciation in the area (Ollier et al., 1988). To the south-east of the main Kimberley massif, separate groups of hills, each with its own set of accordant summits sloping south-east but at different levels along strike, probably reflect the topographic effect of faulting along the Halls Creek Mobile Zone during the formation and subsequent deformation of the Bonaparte and Ord Basins.

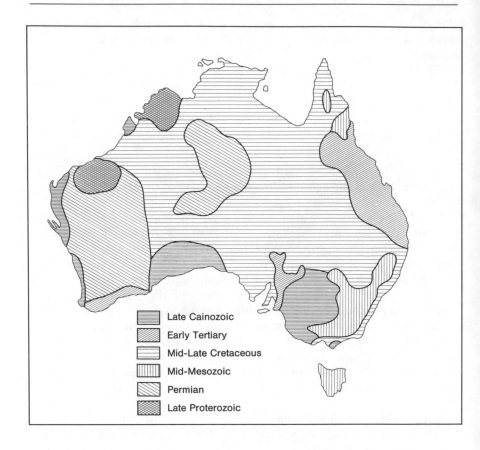

Figure 7.17 Age of development of drainage in different regions of Australia. Map prepared by G. Wilford.

The Hamersley–Pilbara area major relief features are, like those of the Kimberley, developed in highly resistant sedimentary sequences that have accordant summits in the form of broad domes. Major drainage is towards the west-north-west and the almost linear Fortescue follows the outcrop of a very gently dipping Early Proterozoic carbonate unit, suggesting the area has been tectonically quiescent for a considerable time. Permian glacial erosion was probably a major factor in modifying the shapes of some valleys as diamictites, and ice-polished surfaces are present and large quantities of fluvioglacial sediments were shed into the adjacent Canning and Carnarvon Basins. The accordant hill summit pattern (Hammersley Surface) — reminiscent of that in central Australia — could be a smooth surface moulded in part at the outset of the Late Proterozoic glaciations and uplifted isostatically since as erosion has carved the present valleys.

2. Permian drainage

In central Australia much of the relief probably dates from the Carbo-Permian glaciation as Late Permian pollen is present in alluvium close to a range in the Hermannsberg area. Contours on accordant summits show two high areas coincident with the watersheds of the present, presumably superimposed, drainage. Uplift consequent on the removal of softer strata during the Carbo-Permian glaciation may have given rise to the present relief in the Alice Springs area where some of the major ranges are characterized by flat summits. These summits could be remnants of a pre-glacial erosion surface of low relief.

3. Mesozoic drainage

Drainage in the south-east highlands and the upland parts of Cape York Peninsula probably developed throughout the Mesozoic as, prior to Late Cretaceous uplift, a considerable proportion of the two areas had been reduced to broad, undulating lowlands. In Tasmania it seems likely that the main drainage (though considerably modified by Tertiary faulting and Pleistocene glaciation) may date from Mid-to Late Jurassic times as the intrusion of thick dolerite sheets and any eruptives, since removed, will probably have disrupted any preexisting drainage.

4. Mid-Late Cretaceous drainage

For the low-lying parts of the continent the drainage was initiated by the withdrawal of the sea during the Albian and, in the east, affected by the continued downwarping of the Karumba and Eromanga Basin depocentres. Drainage across the latter was modified by local tectonism in Mid-Tertiary times. Warping in the northern part of the Northern Territory resulted in radial and rectilinear patterns of drainage being superimposed on Precambrian rocks from a thin Early Cretaceous cover.

Throughout large areas of the low lying 'palaeochannel country' of Western Australia, South Australia and the Northern Territory, outcrop patterns and the distribution of regolith features suggest that widespread inversion of drainage probably took place, mainly in the Late Cretaceous, most of the 'palaeochannels' dating from that time. The deposits of the earlier rivers remain as sand and gravel plateaux, silcreted in places.

5. Early Tertiary drainage

Drainage into the Southern Ocean along the southern part of the Yilgarn area was initiated following the withdrawal of the sea in the Late Eocene. Drainage in the Mount Lofty/Flinders Ranges region is likely to have been mainly initiated in Paleocene or earlier time and to have been periodically rejuvenated throughout the Cainozoic. Drainage along the eastern highlands

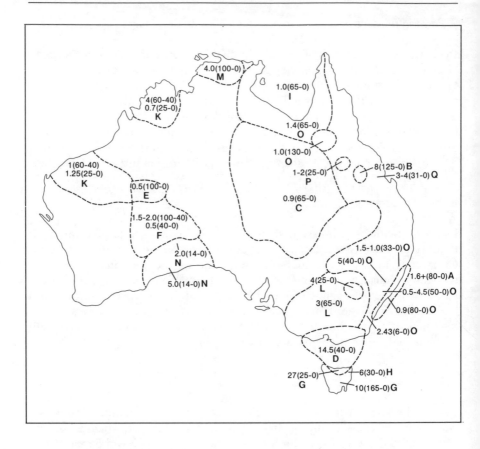

Figure 7.18 Selected Mesozoic-Cenozoic erosion rates in Bubnoff units (Metres per million years). Time ranges are given in brackets. Letters refer to references below. Map prepared by G. Wilford.

A. Based on estimate of rock removed seaward of coastal scarp along NSW seaboard, sediment accumulated beneath continental shelf and rise since uplift of Eastern Highlands 80 million years ago (C.D. Ollier, pers. comm.)

B. Rate of unroofing of 125 million year old Bundanna Granodiorite (Galloway, 1987)

C. Lake Eyre Basin contains an estimated 18,000 km^3 of Cenozoic sediment from a source area of 500,000 km^2.

D. Post Late Eocene sediments in Bass Strait Basins are estimated to total about 60,000 km^3 from source areas of 55,000 km^2 (Victoria) and 45,000 km^2 (Tasmania).

E. Estimated on removal of 50 m of flat-lying Cretaceous and Permian strata from the north-western Officer Basin since the Mid-Cretaceous.

F. Van de Graaff, 1981.

G. Sutherland, 1977.

H. Sutherland and Wellman, 1986.

I. Karumba Basin sediments are estimated to average 150 m thick from source area of 475,000 km^2.

J. Lake George Basin averages 80 m sediment from catchment of 730 km^2.

K. Estimates of clastics volumes in the adjacent predominantly carbonate North-West Shelf sequence.
L. Bishop, 1985.
M. Chappell (pers. comm. to P. Wellman, 1987).
N. Lowry and Jennings, 1974.
O. Wellman, 1987.
P. Galloway, 1987. Rate of removal of deep weathering profile.
Q. Sutherland 1977.

between Brisbane and Townsville which probably mainly flowed west or south-west, was diverted coastwards as a result of faulting and downwarping parallel to the coast in Early Eocene time.

6. Late Cainozoic drainage

Drainage over the Perth Basin has been greatly influenced by Late Cainozoic eustatic incursions, that over the Carnarvon Basin by post-Miocene tectonic movements that are continuing. Drainage has been sub-surface beneath the Nullarbor Plain area since the retreat of the sea from the area in Mid-Miocene times. Drainage in the central and western parts of the Murray Basin followed retreat of the sea in Mid-Miocene times and the draining of Lake Bungunnia in the Plio-Pleistocene. In western Victoria some drainage appears to be consequent on doming (Dundas Plateau), probably the result of sub-basaltic intrusion.

A map showing rates of erosion can also be prepared, and again Wilford (pers. comm.) has made a map for Australia (Figure 7.18). It cannot be used to date surfaces directly, but gives some indication of how likely it may be for remnants of old surfaces to survive. The lower the erosion rate, the more likely they are to survive, and the higher the erosion rate, the less probable long-term survival becomes. The figures for such a regional map must be very general averages, however, and survival of small remnants is possible even where average rates of erosion may suggest otherwise.

One of the most significant features of this map is the time span utilised in deriving the rate of erosion. The results provide much food for thought, and it would be revealing to compare them with modern erosion rates. All three maps show that long-term landscape evolution is not a conceptual oddity, but a geomorphic necessity that can be expressed in quantitative maps.

Chapter 8
Plate tectonics and continental margins

The most obvious single feature of the earth's surface is the division into continents and oceans. This is not an accidental distribution, with water filling hollows on an irregular surface, but a reflection of a fundamental geological difference between continents and oceans.

Continents are predominantly of granitic composition, and the oceans are floored by basalt. The continents, rich in silica and aluminium, are a sial layer; the sea-floors, rich in silica and magnesium, are a sima layer. The sialic continents, being less dense, 'float' on a sima layer, for basalt is also present at depth under continents. Not only are the continents different from oceans, but some of them have shapes which suggest that they may be fitted together like pieces of a jig-saw puzzle. This led to the hypothesis of continental drift. Today the idea is widely accepted, under the name of plate tectonics.

Plate tectonics

Most modern geology textbooks have plate tectonics as a ruling theory, and there are many good accounts including Cox and Hart (1986), Condie (1989), and Howell (1989). As Howell (1989) wrote: 'Plate tectonics, largely illuminated by geophysicists with an oceanic data base, provides the continental geologist with a powerful conceptual tool.' There is a notable lack of any mention of geomorphology in this statement, but any understanding of large-scale landforms must include tectonic geomorphology, and geomorphology does put constraints on tectonic hypotheses (Ollier, 1981).

It was found that there are major lines in the oceans, such as the Mid-Atlantic Ridge, where new sea-floor is generated. Older sea-floor moves aside, so the newest strip is bounded on each side by a strip of older sea-floor, outside which are strips that are still older. Sea-floor spreading is symmetrical, and is recorded in magnetic signatures. Large parts of the ocean have now been mapped, and maps showing the age of the sea-floors of the world have been published. There is no ocean floor anywhere on earth that is older than about 200 million years (Jurassic). Of course, any understanding of the physiography of the ocean floor must take into account its age and long-term evolution.

The continents move aside as the sea-floor spreads. As the North Atlantic Ocean was created by sea-floor spreading, Europe and America drifted apart; as the Southern Ocean was created by sea-floor spreading, Australia moved northwards away from Antarctica.

If the Atlantic Ocean is growing, the Pacific Ocean should be shrinking as Eurasia and the Americas drift towards the Pacific from both sides. Yet the Pacific is also spreading, from the Pacific Rise. The space problem is solved by the process of subduction, in which some of the earth's crust is lost by sliding under another part of the crust when two moving pieces of crust collide. In plate tectonic theory the earth is seen to consist of a number of 'plates' bounded by spreading sites and subduction sites (Figure 8.1). Some plates, such as the Pacific Plate, consist of ocean only; others, such as the South American Plate, consist of both continental and oceanic crust. Subduction sites are collision sites, and the bounding land is called an 'active margin.' Continental margins with no subduction are called 'passive margins.'

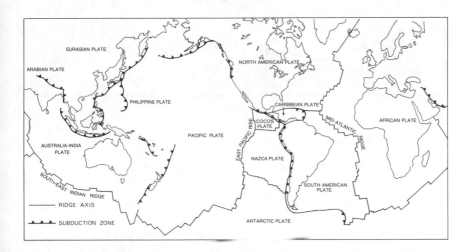

Figure 8.1 The major plates, ridges and subduction zones of plate tectonics. (Source: C.D. Ollier, *Tectonics and Landforms*, Longman, 1981)

Another way of thinking of the plates is like the plates of a turtle-shell. If a continent is bounded on all sides by spreading sites, sea-floor spreading will add more and more strips of sea-floor to the edge of the plate, which will therefore grow. Antarctica more or less fits this description (Figure 8.2). In fact all the continents grow to have an enlarged shape which is a kind of caricature of the continental landmass. If all the plates were growing in all directions the world would have to be expanding, a hypothesis propounded with a wealth of detail by Carey (1976, 1988). But many of the plates have expansion on only part of their perimeter, and on the non-expanding edge it is presumed that subduction is taking place.

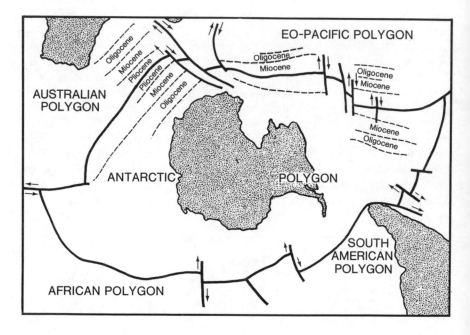

Figure 8.2 The Antarctic Plate. The plate has grown by accretion of new sea–floor around the Antarctic continent. The perimeter of the plate consisting mainly of spreading sites has grown, since the spreading sites originated as rifts adjacent to the continent. The mid-ocean ridges therefore also move over the globe, away from the continent on opposite sides. These passive, responsive ridges with their volcanoes migrate over the surface of the earth. (Source: C.D. Ollier, *Tectonics and Landforms*, Longman 1981)

In the days when the simple theory of continental drift held sway, the reassembled continents were often united to form two super-continents, Gondwanaland (the southern continents) and Laurasia (the northern continents). We now know that Gondwanaland and Laurasia were united, so the continents of today have been created by the breakup of a huge former continent called Pangaea (sometimes spelt Pangea). The breakup of continents is continuing, with rifts slowly evolving into sea-floor-spreading sites. Geomorphology is concerned with the nature of the breakup (Chapter 12), and the long-term evolution of passive and active margins.

Rhombochasms

If a sea-floor-spreading site simply grows wider, a roughly rectangular piece of new oceanic crust will be formed which is called a rhombochasm. The Red Sea is a narrow rhombochasm, and the North Atlantic is a wide rhombochasm. A small rhombochasm is the Dead Sea, which involved strike-slip movement and not just widening (Figure 8.3).

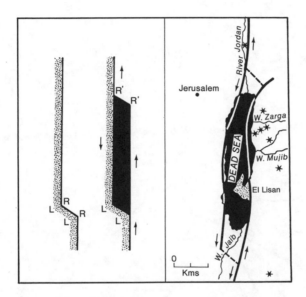

Figure 8.3 The Dead Sea rhombochasm. Left, diagrammatic, and right more realistic. Note the lack of deltas off Wadi Zarga and Wadi Mujib, and the delta at El Lisan with no apparent feeder. If the delta was built by these wadis in former time there has been about 40 km of movement since they parted. Using such evidence Quennell (1959) found that lateral movement in the region anmounted to 107 km, of which 62 km was Early Miocene to Early Pliocene, and a second phase of 45 km from Middle Pleistocene to present. (Source: C.D. Ollier, *Tectonics and Landforms*, Longman 1981).

Sphenochasms

Instead of symmetrical sea-floor spreading causing a rectangular basin of new sea-floor (a rhombochasm), a wedge-shaped area of sea-floor can be formed, a sphenochasm. Possible sphenochasms are the Bay of Biscay (Figure 8.4), and the Gulf of Arabia (Figure 8.5). Note that compressive effects opposite the sphenochasm may cause a sphenopeism (a wedge-shaped zone of compression like the Pyrenees) or an orocline (a mountain range bent in plan, like the Baluchistan orocline). The North Fiji Basin appears to have started as a

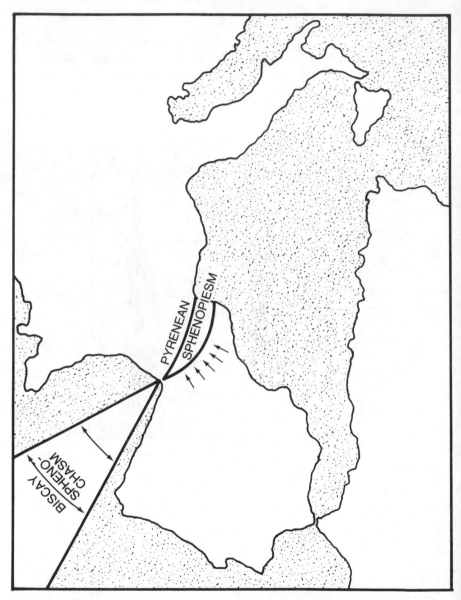

Figure 8.4 The Bay of Biscay sphenochasm and the Pyrenean Sphenopeism (after Carey, 1958). (Source: C.D. Ollier, *Tectonics and Landforms*, Longman 1981).

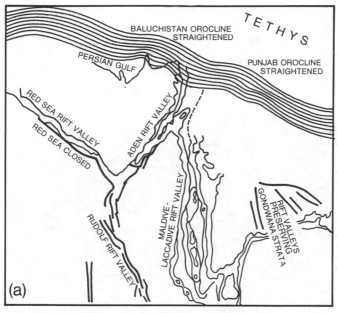

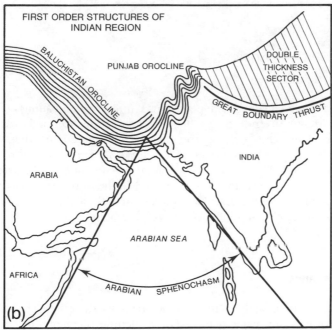

Figure 8.5 The Arabian Sea sphenochasm and Baluchistan orocline (after Carey, 1958). (Source: C.D. Ollier, *Tectonics and Landforms*, Longman 1981)

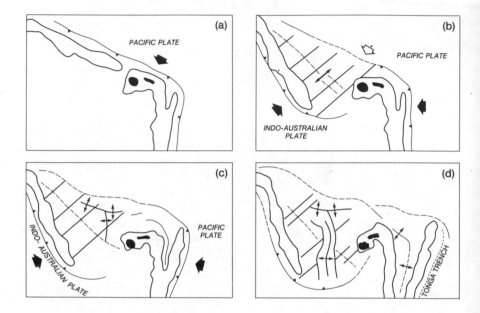

Figure 8.6 Evolution of the North Fiji Basin (after Auzende et al. 1988.)
a. 10 million years ago
b. 8 to 3 million years ago
c. 3 to 0.7 million years ago
d. 0.7 million years ago to present.
The black islands represent Fiji

sphenochasm which later developed a new symmetrical spreading site (Figure 8.6) according to Auzende et al. (1988). The spreading occurred over the past 10 million years, and this scenario may be considered together with the evidence (p.115) for the rotation of Viti Levu, the main island of the Fiji group. Locardi (1988) suggests that a unique arc (sphenopeizm) existed in the Apennine region of Italy from Oligocene to Middle Miocene times as a result of the rotation of the Sardinia–Corsica continental block, caused in turn by opening of the Ligurian Sphenochasm (Figure 8.7).

Exotic terranes

A later complication on the plate tectonic theme was that there are many small units as well as the major plates. These include fragments of continents, fragments of continental margins, fragments of volcanic arcs and fragments of ocean basins (Howell, 1989). These fragments are known as 'terranes'. Terranes begin either by being rifted off a continent, or growing intact in an ocean setting. Such terranes drift around and may collide or amalgamate. A

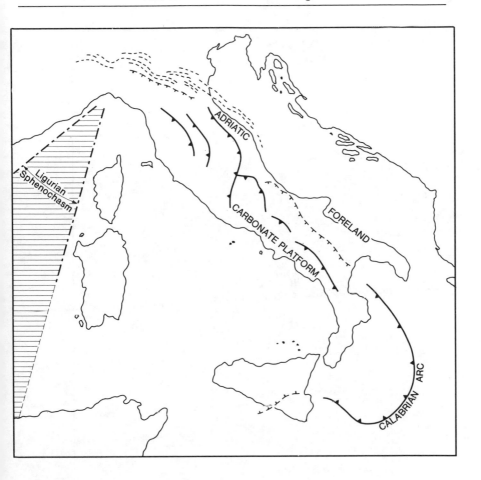

Figure 8.7 The Ligurian sphenochasm which opened between Oligocene and Middle
Miocene times, with the formation of the associated Apennine arcs (after Locardi, 1988).

terrane may be plastered on to the edge of a continent, and will have quite a
different stratigraphy from the rocks with which it comes in contact. It is
therefore known as an exotic terrane, allochthonous (far-travelled) terrane, a
morphostratigraphic terrane, or (if there is doubt) a suspect terrane. Terrane
maps have been produced for many areas, especially of the west coast of North
America (Figure 8.8.).

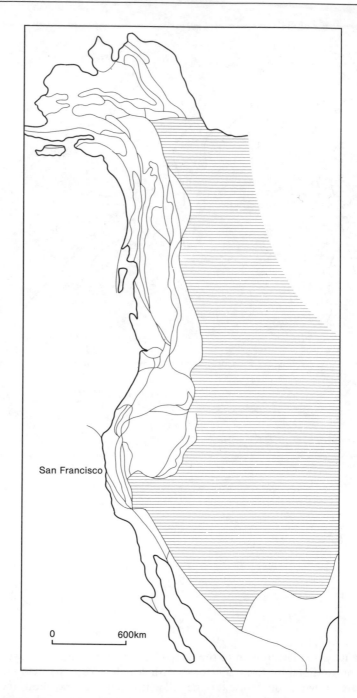

Figure 8.8 Exotic terranes of North America (after Coney et al. 1980)

Equally important, there are places where exotic terranes appear to be excluded. In the Andes, according to Dalziel (1987)

There is no evidence that collision of exotic terranes were of major significance in the building of the Andes.... Indeed the igneous and metamorphic complexes cited in the literature as possibly exotic were largely incorporated in the S American segment of the Gondwanaland supercontinent before the Middle Triassic–Late Jurassic initiation of the present mountain range.

Examples of exotic terranes

The Paleocene Chuckanut Formation is deposited on a collage of terranes in Washington and British Columbia, providing a minimum age for the accretion of several terranes in this part of North America (Howell, 1989). The South Island of New Zealand is composed of several terranes of different composition, including Palaeozoic continental strata, Mesozoic volcanic island arcs, and oceanic material. The terranes did not amalgamate until the middle of the Cretaceous, but palaeomagnetic studies of Triassic and Jurassic strata indicate deposition at about 66g S, which is where New Zealand would fit by simply closing the Tasman Sea and Southern Ocean (Howell, 1989).

In terrane theory, Taiwan results from the collision of a volcanic arc which moved westwards from an origin on the Philippine plate and collided with the continental margin of China. The Coastal Range along the east part of Taiwan is a remnant of the old arc, and the Central Range is a deformed part of the China margin (Howell, 1989). In this instance the morphogenetic stage of mountain building is directly related to terrane accretion, which is dated at about 6 million years.

Papua New Guinea consists of a large number of exotic terranes according to Pigram and Davies (1987), who identified thirty-two terranes in the north of the island and have worked out the history of docking of terranes through the Tertiary (Figure 8.9). Each terrane has a unique stratigraphic and structural history, reflecting widely different conditions such as continental basement or nearshore sediment (Kemum terrane), deep water chert and carbonate terrane (Port Moresby terrane) and ophiolite basement terrane (Waigeo terrane) Many of the terranes are basins of Miocene age containing 3–7 km of turbidite sediment. According to Pigram and Davies the terranes were detached from eastern Gondwana in the Mesozoic and rejoined the craton by docking in the Mid-Tertiary.

Problems associated with exotic terranes have already been mentioned briefly in Chapter 4 for river patterns and histories on exotic terranes would have no genetic relationship to those on the main continent. The same would apply to structural landforms, depositional basins and in fact all geomorphic features. The same thing is likely to happen to a lesser extent with strike-slip faulting of large dimensions. This casts doubts on the time-honoured way of

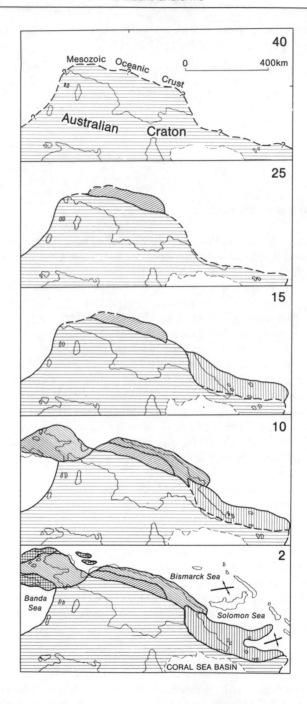

Figure 8.9 The accretion history of exotic terranes to make New Guinea (after Pigram and Davies, 1987). The numbers show millions of years ago.

reasoning along 'geotraverses', by drawing sections along a line and trying to understand the relationships between the different palaeogographic units. This is a necessary and unavoidable exercise, but the method (in the words of Trumpy, 1983) 'neglects the lateral movements between continental and oceanic slivers, which may lead to juxtaposition of unrelated palaeogeographic units, to cutting out of palaeogeographic belts in one geotraverse and duplicating it in another'.

Rotation of plates

Palaeomagnetism can be used to detect rotation of plates or crustal fragments, either in the ocean or on land. In Fiji analysis of marls shows the rotation of the island of Viti Levu by 31 degrees over the last 6 million years (Cassie, 1978). This rotation would have secondary geomorphic effects by changing relations between the island and prevailing winds, climatic regions, ocean swell and so on. Further work by Inokuchi, Yaskawa and Rodda (pers. comm.) showed (1) a clockwise rotation of about 45 degrees during late Oligocene, (2) an anticlockwise rotation of about 75 degrees since the late Miocene. Clockwise rotation was associated with the opening of the South Fiji Basin and the anticlockwise rotation was related to the opening of the North Fiji Basin. The rotation is consistent with what is known of sea-floor spreading in the North Fiji Basin (Figure 8.6).

Rotation can also take place in a continental setting. Li et al. (1990) report that Mid-Miocene dykes and flows 14 to 17 million years old at three localities in northern Nevada rift show that a block or blocks have rotated about 19 degrees relative to stable North America. Palaeomagnetic studies of rocks from Chihuahua in northern Mexico suggest that a counterclockwise rotation relative to stable North America occurred during Early Tertiary time (Urrutia-Fucugauchi, 1981). Oligocene volcanic rock from the Cascade Range in southern Washington indicate a counterclockwise rotation of about 33 degrees relative to stable North America (Bates et al., 1981). Palaeomagnetic data from the Cascade Range in central Washington to northern California indicate that a coherent block has undergone about 27 degrees of clockwise rotation in the past 25 million years (Magill and Cox, 1981). It seems rotation occurred in two phases, the first being an Eocene rotation that accompanied the accretion of the Oregon Coast Range to the continent, and the second phase being a rotation associated with post-Early Miocene extension in the Basin and Range Province.

Continental margins

There are two kinds of continental margins — active and passive. At any margin there are two aspects of interest in long-term geomorphology: the evolution of the landward part of the margin, and the sedimentation offshore. Clearly the two must be related, but to tell the story simply this chapter will concentrate on geomorphology and the sedimentology will largely be placed in the next chapter, Oceans and Coasts.

Passive margins

New continental margins and coasts came into existence with the breakup of super-continents. The edge of the siallic continent is roughly at the continental shelf, and the coastline is a rather accidental line where the sea happens to lap onto the continental mass. The Atlantic opened in the Jurassic, so no coastal features can pre-date that. Some major features have grown up since, such as the Niger delta, which has to be 'removed' to give a good continental fit between South America and Africa in plate tectonic reconstructions.

The breakup of Gondwanaland occurred at different times in different places. Australia separated from Antarctica about 55 million years ago, although initial rifting may have taken place about 100 million years ago. The Tasman Sea, between Australia and New Zealand was created by sea-floor spreading that started about 80 million years ago.

Passive margins undergo two phases of development: a rifting phase, generally thought to be characterized by thinning of the lithosphere with probable associated uplift on the landward side, and a subsidence phase, possibly later, where the margin subsides as it cools and is loaded with sediment along the newly formed continental edge.

Many passive continental margins are bordered onshore by asymmetrical marginal bulges (*Randschwellen* in German, or *bourrelets marginaux* in French). These have been studied, especially in the northern hemisphere (Godard, 1982; Battiau-Queney, 1984; Peulvast, 1988, Le Coeur, 1988) and many seem to be related to the opening of the Atlantic. An example from the southern hemisphere is shown in Figure 9.1, with what King called 'rim highlands'.

There are many possible causes for the formation of marginal swells, as for uplift in general, and a list is provided in Table 8.1. Theories as different as isostatic recovery after erosion of earlier mountains, underplating of the crust by light material of unknown nature, and response to thermal phenomena associated with earlier rifting should all produce different effects, probably at different times. If the hypothesis is sufficiently developed to allow prediction, and the age of geomorphic development can be determined, then the hypothesis can be tested. Thus geomorphic features of passive margins should be able to put constrains on tectonic theory, and perhaps even afford crucial tests.

For example, Smith (1982) proposed a model for uplift of southern Africa based on migration of the African continent over a source of heat parallel to the east coast. 'those areas affected by plateau uplift now overlie or have passed across the former positions of the oceanic ridges that separated Madagascar and India from Africa.' But the model is based on perceived asymmetry and accounts for uplift only in the east of the region. In reality the marginal swell

Table 8.1 Some possible causes of formation of marginal swells on passive continental margins.

1.   Passage of the continental margin over a zone of anomalous high heat flow (eg. Smith, 1982; Karner and Weissel, 1984).

2.   Thermal expansion of a mantle plume beneath the continental margin (eg. Cox, 1989).

3.   Isostatic response to erosion of the continental margin, especially by escarpment retreat (eg. King, 1955).

4.   Uplift to compensate for subsidence offshore, caused in turn by loading of deposited sediments (eg. Gilchrist and Summerfield, 1990). Mechanisms 3 and 4 usually work together.

5.   Underplating of the continental margin by masses of light rock (eg. Wellman, 1987).

6.   Intrusion of large amounts of igneous rock.

7.   Delayed response to erosion of earlier orogenic belts (eg. Lambeck and Stephenson, 1988).

8.   Subsidence of basins on each side of an originally high continental margin (eg. Ollier, 1992).

Some other possible mechanisms are listed in Table 13.1

and the Great Escarpment in southern Africa make a complete horseshoe around the continental margin. Africa could not be drifting in all directions over conveniently located heat sources.

Smith has a similar explanation for south-east Australia: 'Southeast Australia moved onto the former ridge position in the Tasman Sea shortly after it had opened between 76 and 56 m.y. If the Australian lithosphere is similar to Africa I expect strong uplift beginning about 16 m.y. ago.' Nowadays no workers on the Southern Highlands of Australia would believe that uplift was so young (Wellman, 1987, 1988; Bishop, 1986, 1988; Ollier, 1992). Furthermore there is a unity of the eastern highlands of Australia, but the

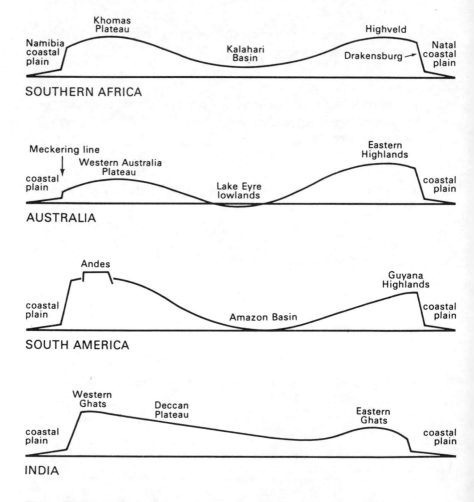

Figure 8.10 Great Escarpments and marginal swells. Cross–sections across several continents to show the general pattern of marginal swells with a basin in the centre and Great Escarpments facing the ocean.

entire continental margin does not have the same relationship to spreading sites.

Great Escarpments

Many continents exhibit a very spectacular landform along their passive margins: the Great Escarpment. This is a landform on the grand scale, thousands of kilometres long, and often around a thousand metres high. Despite their size, until about the past decade the study of Great Escarpments was strangely neglected. In southern Africa the feature is often called the Great Escarpment in ordinary talk, and has long been a subject of debate. In India the scarp has long been recognized, but the name Western Ghats did not convey the nature of the geomorphic feature to those who have not seen it. Many of the world's great mountain 'ranges' are in fact plateaux bounded on one side by a Great Escarpment: the Drakenberg, the Transantarctic Mountains, the Snowy Mountain and the Serro do Mar are just a few examples. In fact the large number of local names obscured the unity of the Great Escarpment, so that in Australia, for instance, the unity was reported only in 1982. The possible morphotectonic similarities between several continents are shown in Figure 8.10.

Great Escarpments tend to run parallel to the coast, and they separate a high plateau from a coastal plain (Figure 8.11). In reality they separate two regions of vastly different geomorphology: the tablelands with many palaeoforms, low relief and slow process rates, and the coastal zone with few palaeoforms, moderate relief and rapid process rates. This functional relationship can be discovered even when the escarpment is not very high: the Meckering Line in Western Australia bounds a plateau with ancient drainage lines, but is not a very high escarpment.

The top of the Great Escarpment can be very abrupt (Figure 8.12). The

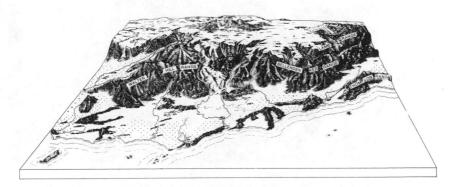

Figure 8.11 The Great Escarpment near Innisfail, south of Cairns, Queensland. In the centre is a lava flow which has poured over the escarpment and now has twin lateral streams.

Figure 8.12 Air photos of the Great Escarpment near Armidale, New South Wales (Photo: C.D. Ollier)

Figure 8.13 The Great Escarpment at Wollomombi Falls, northern New South Wales. The falls are about 200 m high. Note the minor valleys preserved at the top of the escarpment, like valleys truncated by a marine cliff. (Photo: C.D. Ollier)

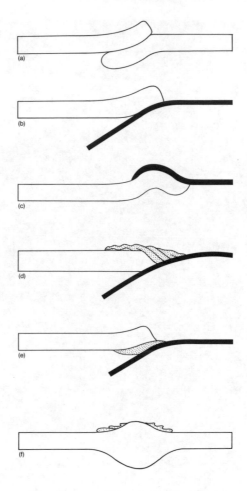

Figure 8.14 Some possible effects at collision sites in plate tectonic theory.
a. Continent–continent collision (Himalayan type).
b. Continent–ocean collision, with uplift of the continent and subduction of the ocean floor (Andes type).
c. Obduction of sea–floor over the continent, with later isostatic rise to form mountains (Cyprus type).
d. Thrusting of marginal sediments on to the continental plate with foreland-folding at the front and thrust-faulting at the rear (Appalachian type).
e. Subduction of ocean plate and sedimentary wedge. Later melting may produce granite intrusions and andesitic volcanism.
f. Continent–continent collision leading to crustal thickening, possibly accompanied by gravity sliding of rocks near the surface.
(Source: C.D. Ollier, *Tectonics and Landforms*, Longman 1981)

Great Escarpments are undoubtedly erosional, despite the tendency for writers on plate tectonics to draw cross-sections of passive margins showing only normal faults downthrown on the ocean side. The mechanism of parallel slope retreat seems to be responsible, following drainage lines to make great gorges. Most of the world's large waterfalls are found where rivers cross the Great Escarpment. In places the escarpment retreats as a wall, leaving old shallow valleys perched at the top of steep cliffs (Figure 8.13).

Active continental margins

In plate tectonic theory, most of the action takes place at what are called active margins, or collision sites, where plates are converging. A great many responses to collision are possible, and collision is used to explain uplift (building mountains), subsidence (making deep sea trenches), the generation of folded strata by compression, or the opening of back-arc basins by tension. Some of the possibilities are shown in Figure 8.14. Further discussion of active margins will be deferred until Chapter 13.

Chapter 9
Oceans and coasts

Deposition on passive continental margins and breakup unconformities

After the breakup and creation of a new continental margin, the continent is warped, often creating an upwarped rim or marginal swell onshore, and offshore downwarping produces a new sea-bed. Horizontal sediments are laid down on the sea-bed, clearly different from the usually folded, faulted and intruded bedrock beneath. The plane between the young sediments and old bedrock is called the 'breakup unconformity.' Sediments have built up since the inception of the new seafloor, usually to form a sedimentary wedge. This will date from the time of origin of the new continental margin. On the New South Wales coast, for instance, the oldest sediment is likely to be about 80 million years old, for that is when the bordering Tasman Sea was created.

Young ocean basins between continents are rare on the earth, and so are young passive continental margins. Most passive continental margins are old and covered by thick sedimentary sequences that prevent any immediate insight on the early stages of their evolution. In this regard the Gulf of Aden and its African and Arabian continental margins represent a fortunate exception, where Abbate et al. (1986) were able to describe the pre-rift and post-rift setting well exposed along the coast.

The uplift of the Appalachians is attested most clearly around the south end of the range where the Cretaceous and Cainozoic sediments of the Coastal Plain dip a degree or two away from the mountains (and dip less, the younger they are). The margins of the range have gradually been buried under these sediments, beginning in the Jurassic (Rodgers, 1983).

In Georgia, United States, the breakup unconformity is a composite of overlapping weathered surfaces. Along the Fall Line (a narrow zone connecting waterfalls on near parallel rivers, the line between the old rocks of the Piedmont Plateau and the younger rocks of the Atlantic coastal plain), the unconformity is Upper Cretaceous on the west side of Georgia and Middle to Upper Eocene on the east side of the State. Downdip from the Fall Line for about 20 km the unconformity is over metamorphic and igneous rocks. Farther down dip it becomes a Triassic weathering surface over siltstones and basalt (Pickering and Hurst, 1989).

The mass of sediment offshore should correspond with erosion in the coastal zone, and this is often the case. In New South Wales there is especially thick offshore sediment where the Bega Granite has been eroded, suggesting that the granite was already deeply weathered at the time the coast was created. The total amount of erosion matches the total amount of sediment remarkably well (see Figure 3.8).

A hypothesis that sounds almost mechanically necessary is a vertical rotation of the margin: erosion of the landward margin reduces the crustal load and should lead to isostatic uplift, and the deposition of a large wedge of sediment offshore should increase the load and lead to subsidence. The hinge zone between uplifted areas and subsiding areas should be roughly along the coast. King (1955) and Pugh (1955) used this notion in relation to pedimentation of younger landsurfaces and purported to present evidence to support the idea, with tilting of pediplains as predicted in theory and seen in the field (Figure 9.1). The evidence is not very conclusive, and alternatives are now available for South African landscapes (Partridge and Maud, 1987, 1988). Gilchrist and Summerfield (1990) discuss the differential denudation and flexural isostasy of margins and conclude that denudation could cause as much as 600 m of upwarp. In eastern Australia there is no sign of isostatic compensation, in the form of uplifted intermediate surfaces, although the whole region is in isostatic equilibrium (Wellman, 1979). The nature of the east Australian coastal zone is shown in Figure 3.8. An account of all the Australian continental margins is provided by Falvey and Mutter (1981).

Figure 9.1 The isostatic response of a continental margin to erosion of the continental margin and loading by sedimentation on the offshore side of the coastal hinge-zone, in southern Africa. According to King, maximum uplift took place in a zone from 150 to 250 km inland, and the surface sloped from this zone seaward on one side and towards a continental depression (in this case the ancestral Kalahari) on the other. Erosion surfaces are tilted as shown. (After King, 1955.)

Deposition on active continental margins

Deposition on active continental margins should be more complicated than that on passive margins, and it probably is. On passive margins sediment simply accumulates, but on active margins, often bounded by narrow continental shelves and deep trenches, sediment can be lost to the deep sea. In plate tectonic theory the sedimentary prism can also be subducted under the continent, scraped off on to it, or carried away by strike-slip faulting or on an exotic terrane. Several lines of evidence show the story is not as easy as the simple plate tectonic theory might suggest. In many trenches the sediments are horizontal and undisturbed: numerous trenches have many normal faults and graben indicating tension rather than compression, and some trenches are empty. Scholl and Marlow (1974) noted that the Chile trench, for example, is almost empty. They were perplexed that there was no evidence for subduction or off-scraping of trench deposits, and in response to the suggestion that the evidence has itself vanished down the subduction zone, they add, 'You can't have your trench and eat it too.'

To relate ocean features to terrestrial geomorphology, in the Peru–Chile trench between 36g and 42gS, the estimated volume of Pleistocene clastic debris derived from the continent requires an average erosion rate over the past 1.2 million years of about 500 B. This is the average figure for erosion of highlands, and implies a good balance between the volume eroded from the land and the volume of sediment deposited. But if sediment were lost by subduction over this period, the erosion rate on land would have to be significantly increased, giving an erosion rate that is improbably high, but not impossible. Similar calculations have been made for the Aleutian trench and the conclusion seems to be the same (Scholl, Marlow and Cooper, 1977).

On the Pacific coast of the United States, uplift of the Coast and Insular Mountains in the Late Cenozoic was accompanied by fluvial and glacial erosion. Most of the detritus was carried to the Pacific Ocean. Some of the material may form the upper part of the sedimentary prism underlying the continental shelf, but the volume of Plio-Pleistocene sediments on the shelf is much smaller than the volume of rock eroded from the Coast Mountains since the Miocene (Muhs et al., 1987). Much of the material eroded from the mountains probably reached abyssal depths beyond the continental margin.

Sea-level

The coastal zone is old, relating to the breakup of Pangaea and the formation of new land masses, but how old is the actual coast, and can we find traces of older coasts? Throughout the Quaternary sea-level has varied greatly in sympathy with the growth and melting of ice-caps. In glacial periods ice-caps grow, and the storage of water as ice leads to a fall in sea-level. Most of the storage is in Antarctica, but there is a large ice-cap over Greenland and in the past there have been ice-caps over much of North America and northern Eurasia. Studies of ice ages have revealed an increasingly complex scenario, from an initial idea of an 'Ice Age', through the four major glacial periods of Penck and Bruckner in 1911, to modern ideas of at least 27 major coolings in the past 3.5 million years (Hooghiemstra, 1984).

Workers in the 1950s tried to match coastal geomorphic sequences to the then standard Mediterranean sequence and the mythical four ice ages, not only in Europe but in far away places like New Zealand (Pillans, 1990). This stage was followed by more complex stories following the discovery of more and more alleged glaciations. Now it is clear that coastal studies must stand alone, with conclusions derived from coastal evidence itself, and not fitted to inappropriate theoretical schemes.

For many years it was hoped to build up a history of sea-level changes based on allegedly stable areas, but later work showed up the incredible complications of the situation. Not only are there no truly stable sites to use as benchmarks, but the sea itself does not make a perfect sphere, and the weight of water on a transgressive shoreline itself affects local tectonic movement. Furthermore, sea-level is affected systematically by such factors as winds and low pressure systems. It is simply not possible to be too precise about sea-level at any given locality.

Nevertheless, a rough idea of sea-level changes over the last few hundreds of thousands of years has been attained, and is particularly good for the past 20 000 years. The perspective of time makes more distant changes ever more speculative. In brief, the sea-level was roughly where it is today about 125 000 years ago, sank to about 150 m below present level at the height of the last ice age, around 18 000 years ago, and has now reached a high, typically interglacial level.

Around Australia there is also evidence for Tertiary low sea-levels, not surprising when it is realized that glaciation extends back into the Tertiary, but the evidence is largely offshore, as benches well below present sea-level (R.W. Galloway, pers. comm.). Tectonic movements provide variety, as in south-east Australia, where Tertiary sediments are almost entirely offshore along the New South Wales coast, but have been uplifted to become part of the land along the Victorian coast.

The study of offshore sediments by oil companies has led to the construction of sea-level curves for the entire Phanerozoic period (Figure 9.2). These are called Vail curves, after the pioneer worker in this field. Detailed curves have now been constructed back to the start of Triassic times (Haq et al., 1987). The Vail curves have some remarkable properties, revealing things about global geology, and presumably geomorphology, that were previously unsuspected. The most difficult to explain is the saw tooth nature of the curve, with very sudden falls in sea-level followed by a long, slow rise. In the Quaternary this might be explained by very rapid growth of ice sheets and subsequent slow melting, but the same pattern is found in parts of earth history when the world was virtually free of ice. Another explanation relates to the growth and decline of mid-ocean ridges, and yet another is that earth expansion is accompanied by production of new ocean water, and the changing level reflects the balance between expansion and water production (Carey, 1988). The sudden falls of sea-level should of course mean sudden falls in base level of the rivers, with world-wide periods of incision.

In tectonically stable coasts, present-day coastal landforms are young and

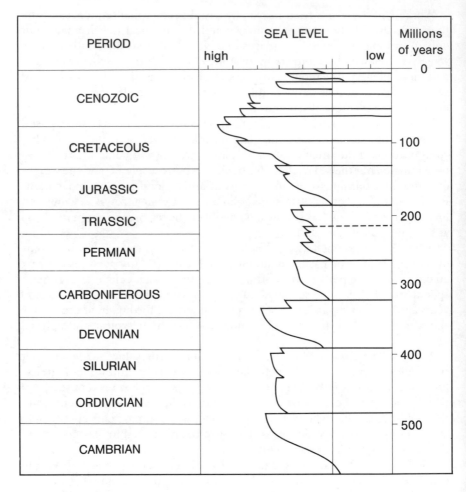

PERIOD	SEA LEVEL	Millions of years
	high low	0
CENOZOIC		
CRETACEOUS		100
JURASSIC		
TRIASSIC		200
PERMIAN		
CARBONIFEROUS		300
DEVONIAN		
SILURIAN		400
ORDIVICIAN		
		500
CAMBRIAN		

Figure 9.2 Vail curves for the Phanerozoic, greatly simplified. Some features to note are the great increase in complexity in the late Cenozoic, the very high levels associated with the Cretaceous flood, and the nature of the curve with very rapid falls in sea level followed by gradual increase. The cause of this kind of curve is not yet generally agreed.

even ephemeral. Complications in interpretation may arise when the present sea-level is almost coincident with that of the last interglacial high sea-level, 125 000 years ago. It is not always clear what coastal features are forming now at the coast, and what are inherited.

In regions of tectonic uplift a greater historic sequence is preserved. Sequences of terraces may be produced. These do not usually reflect a jerky uplift, but rather the algebraic sum of a constant rise of the land against up-and-down movements of the sea. In New Guinea the terraces (Figure 9.3) go back to pre-Quaternary times (Chappell, 1974); in New Zealand terraces

Figure 9.3 Raised coral terraces of the Huon Peninsula, Papua New Guinea.

have been traced to heights of over 1 000 m, and the higher ones, though not directly dated, are probably pre-Quaternary (Bull and Cooper, 1986). Pillans (1990) cautions against the assignments of marine terrace ages based solely on height, as tests show it is unreliable in the absence of other dating methods. Similar high terraces are found in many other parts of the world, including Japan (Ota, 1985).

On a longer time-scale old coastlines can be traced back into the Pliocene, Miocene and even Eocene by preserved beach ridges and associated dunes (p.17). Even older shorelines can be traced by detailed mapping. Lillegraven and Ostresh (1990) have produced a series of maps showing the position of thirty-three Late Cretaceous shorelines in the Western Interior Sea of the United States. These may not be directly expressed in the topography, but they certainly indicated the initiation of terrestrial geomorphology after the sea had retreated, and give a good grasp of the age of landforms in the area. In Georgia, United States, sea-levels fluctuated from an Upper Cretaceous high near 260 m to a mid-Tertiary low of about 60 m above the present sea-level (Pickering and Hurst, 1989).

When the Mediterranean dried up

The Mediterranean can be regarded as the remains of the ancient Tethys Sea which once stretched from the Mediterranean across to a basin which became the Himalayas. At that stage there was free passage of water from the Mediterranean to the Atlantic or the Indian Oceans. Tectonic processes cut off the Mediterranean link to the Indian Ocean, and almost closed the Spanish end of the sea so that communication of oceans was intermittent. When the Mediterranean basin was finally cut off, the inflow of rivers was insufficient to keep it full, and evaporation dried up even to the bottom of the 3 km - deep basin. The same could happen today: if the Gibraltar Strait were closed the Mediterranean would dry up in about 1000 years.

The main evidence for this story is the great thickness of evaporites found in deep sea cores in the Mediterranean, together with faunal evidence. When the opening to the Atlantic became permanent there was a major change of fauna which marks the beginning of the Pliocene, 5.5 million years ago (Hsu, 1972). There would have been remarkable landforms when the Mediterranean started to fill, with a huge waterfall at the Straits of Gibraltar where the Atlantic overflowed into the basin.

The features of the Mediterranean deep-sea floor are not so interesting geomorphologically as the secondary effects brought about by the exceptionally low base level. Major rivers cut down to a base well below any normally conceived base level. The Nile was one. Drilling for the High Aswan Dam revealed a gorge 290 m deep, with almost vertical sides, now filled with sediment (Figure 9.4). Furthermore the bottom 150 m of sediment were marine Pliocene deposits. This canyon is 1200 km upstream from the mouth of the Nile.

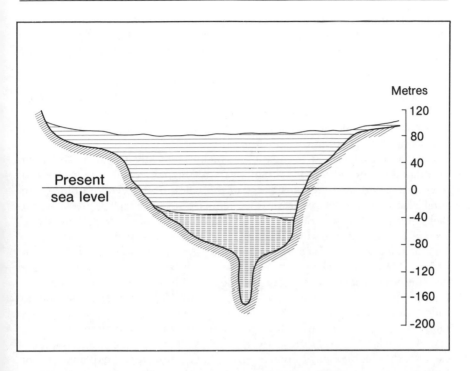

Figure 9.4 Cross-section of the Nile valley near the High Aswan Dam, about 1,200 km upstream from the mouth of the Nile. The deep valley could only have been cut when the base level in the Mediterranean was very low, that is when the Mediterranean dried up. The lower sediments are marine Pliocene deposits, the upper sediments are fluvial.

Similar deep canyons had been found by oil explorers in Lybia, but in the intellectual climate of the time no journal would publish the outrageous results (Sullivan, 1974). On the northern side of the Mediterranean the Rhone valley is deeply incised, and beneath the delta reaches a depth of 900 m. Clearly, to understand the geomorphology of major rivers (and much else) around the Mediterranean a long time-scale is required.

Oceanic islands

Apart from a few fragments of old continent which have been left behind, such as Madagascar, the Seychelles and New Caledonia, most oceanic islands are volcanic. Their geomorphic development begins when their parent volcano emerges above the sea. Volcanic rocks can be dated by the K/Ar method. For instance, Tristan da Cunha in the South Atlantic is about 100 000 years old. The sea has attacked the volcano to produce the 600 m cliffs that encircle the island (Figure 10.13), and the rivers have been unable to cut down to sea-level. The nearby Nightingale Island is very much older and most of the volcanic structure is under the sea, the island being an irregular fragment.

Gough Island 200 km to the south is over 2 million years old (Maund et al., 1988), and is only a remnant of a former much bigger volcano. Like Tristan it is bounded by cliffs. All these islands have some younger, smaller eruptions as well as the main, older volcano.

Some volcanic islands are erupted on the Mid-Ocean Ridge, where new sea-floor is being generated. The Icelandic volcanoes are the best example, and Iceland is often regarded as a portion of Mid-Ocean Ridge that has risen above sea-level. As the plates move apart, older volcanoes are found with increasing distance from the spreading site.

Coral islands grow on foundations which are mostly volcanoes. Darwin postulated a sequence of development from fringing reef to barrier reef and atoll, associated with the subsidence of the volcano. By relative uplift some coral islands are now above sea-level, and many have a central depression that looks like an old lagoon. Some have a whole sequence of uplifts, giving rise to a series of terraces around their coasts. But there are alternative explanations for limestone island with central depressions. According to Bourrouilh-Le Jan (1989), 'Darwin and Davis supposed them to be uplifted atolls but they are not. They are pieces of carbonate platform with Pleistocene or Holocene coral rim around a lagoon on a submerged karst morphology.' Other explanations, called 'solutional saucer' explanations by Guilcher, suggest that the morphology results from simple karstic solution of limestone platforms (Ladd and Hoffmeister, 1945; Purdy, 1974). This is an area where geomorphology can make a valuable contribution in sorting out quite different explanations for landforms derived mainly from the study of sedimentology, but it will have to be geomorphology on a long time-scale.

A sequence of oceanic island development, first described by Chubb (1957) goes from newly erupted volcano, through eroded volcano, and then, when the volcano starts to sink, through a sequence of fringing reef, barrier reef and atoll. In the Hawaii Island, for example, the atolls at the western end such as Midway have local fringing reefs. Next come the main islands, all with drowned river valleys, like Pearl Harbour, with up to 400 m of submergence. Kauai is deeply dissected, and Oahu and Molokai are deeply dissected domes. Then comes Hawaii, still active, and beyond this are five seamounts rising 3 to 4 kilometres above the sea-floor, This is very much the sequence that Darwin described in his theory of coral reef formation, and shows as Chubb wrote, 'subsidence is a usual sequel to volcanic action.' As the islands sink, thick sequences of coral are built up. Enawetok Atoll in the tropical Pacific has 1 405 m of late Eocene to early Miocene limestone on a volcanic basement, under about 200 m of Plio-Pleistocene limestone (Saller et al., 1990). The atoll has apparently been stable through all this time, and there is no tilting associated with the subsidence. The ages of the different stages have now been worked out with a precision that would have impressed Darwin. The oldest oceanic islands have a base of Cretaceous age, e.g. Truk Atoll.

Hot-spots and plate movement

The island sequences so well described geomorphically by Chubb can now be explained by the passage of plates over a hot-spot, a plume of heat in the mantle and a possible source of volcanic eruption. If a moving plate passes over a stationary hot-spot there will be periodic eruptions, all in a line marking the direction of movement of the plate, and with a definite sequence of ages that shows the rate of plate movement, as in Hawaii (Figure 9.5).

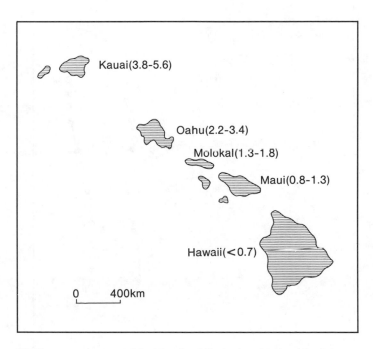

Figure 9.5 The age sequence of the Hawaiian Islands, showing that islands become younger to the south-east consistent with the passage of the Pacific Plate over a hot-spot. The numbers represent the ages of the most recent lavas in millions of years.

A series of submarine volcanoes off the coast of eastern Australia are known as the Tasmantid seamounts, and they form a line over 1300 km long. They have ages ranging from 24 to 6.4 million years (Early to Late Miocene) and show systematic north to south younging of the volcanism (Figure 9.6). They provide strong evidence of a hot-spot track, recording the motion of the Australian plate at an average rate of 67 + − 5 mm/year. This direction and rate is similar to that derived from a study of large central volcanoes on the east Australian continent (Wellman and McDougall, 1974).

If there is a change in direction of plate movement over the hot-spots, there should be a change in direction of the island chains. This is observed in the real world, and the Emperor chain of islands continues the Hawaiian chain, but in a different direction (Figure 9.7). The change occurred 45 million years

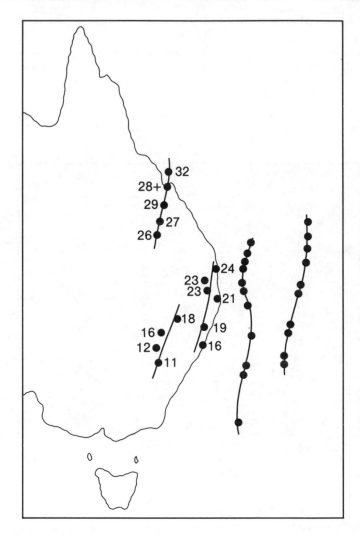

Figure 9.6 Hot-spot traces in eastern Australia and the Tasman Sea. Each dot represents a central-type volcano. The numbers give the average age of the volcanoes in millions of years.

ago (Clague et al., 1975). If there are several hot-spots, there will be several lines of islands with the same direction and the same rate of plate movement revealed by their ages. This is indeed found, and other lines not only have a trend parallel to that of the Hawaiian chain, but also show the change of direction of plate movement (Fig. 9.7). Several authors have related hot-spots to specific reference points in an attempt to work out relative plate movements, but according to Molnar and Stock (1987) hot-spot traces do not define a fixed reference frame.

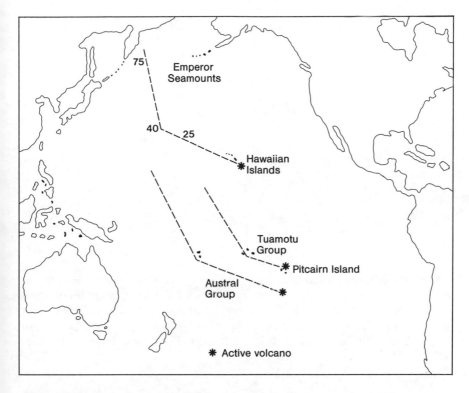

Figure 9.7 Hot-spot traces in the Pacific, showing a change in direction of the hot-spot trace about 40 million years ago.

A warning is sounded by Haggerty et al. (1982), who studied the southern Line Islands in the Pacific. Reef-capped Late Cretaceous volcanoes, 70 to 75 million years old, do not display a regular progression of ages, so this line at least is not a hot-spot trail.

Chapter 10
Volcanoes

Volcanoes differ from most landforms in being made by construction rather than by erosion of older rocks. They are built by spectacular eruptions of fluid lava and fragmentary material (ash, cinders and bombs) known collectively as pyroclastics. The geology and geomorphology of volcanoes are described in many texts, including Francis, 1976; Bullard, 1977; Cotton, 1944; and Ollier, 1988a.

Ages of volcanoes

The age of a volcano can be determined by many methods, including potassium/argon dating, palaeomagnetism, the age of underlying fossils, and others. Some volcanoes have a brief history of eruption, some erupt over a long time, and some have several eruptions of short duration widely spaced in time. Of more importance is the fact that volcanoes enable us to put a time-scale on landscape evolution, both on the volcano itself and in the surroundings. If we see a large mountain which is the remains of a volcano, and that volcano turns out to be 30 million years old, there is no escape from the idea that our landscape history must go back at least 30 million years, and probably much further if we take into account the geomorphology of the basement on which the volcano stands.

Sequence of erosion

At the end of an eruption many volcanoes have a simple conical shape, and erosion later produces a regular sequence of forms through a planeze stage, a residual stage and finally a skeletal stage. Many variations are possible on this theme, and sometimes planezes are still preserved when the plugs are exposed. Nevertheless, the erosion affords a relative dating technique. This

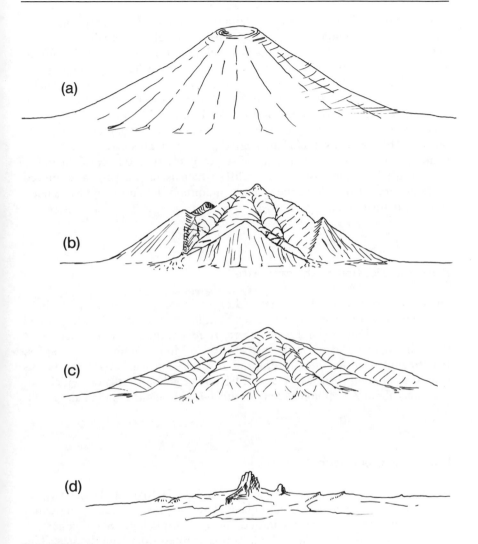

Figure 10.1 Stages in erosion of a volcano.
a. intact volcano
b. planeze stage
c. residual volcano
d. volcano skeleton
(Source: C.D. Ollier, *Volcanoes*, Blackwell, 1988)

was applied by Kear (1957) in New Zealand where it was suggested that stage corresponds to age as follows (Figure 10.1):

Planeze   Mid-Pleistocene to Holocene
Residual   Plio-Pleistocene
Skeletal   Upper Miocene

This mode of erosion can be detected even in some very low-angle volcanoes. A 20 million-year-old volcano near Inverell in Australia, important because of its association with a major sapphire field, exhibits radial drainage and planezes even though the original volcano probably had slopes of only one or two degrees. Obviously it fails completely to fit into the New Zealand age/stage sequence. Other older volcanoes do not seem to show the planeze pattern, such as the Miocene volcanoes of Uganda. Mount Elgon is a huge cone about 50 km across, built of a succession of agglomerate deposits separated by ash layers. Radial drainage is present, but instead of planezes the flanks consist essentially of a series of steps or structural terraces. The easily eroded ash is at the base of each cliff, which is in the vertically jointed agglomerate. Further north the volcanic mountains of Torror and Kadam are simply irregular.

Repeated eruption on the same site

Some volcanoes demonstrate repeated eruption on the same centre. Valleys eroded on an earlier volcanic cone become filled with younger lava flows, and new cones are built up on the old centre. In recent times a classic example is the building of Vesuvius on the site of the old Monte Somma, near Naples, Italy, mainly resulting from the eruption in AD 79. The gap between eruptions may be millions of years, as on Gough Island in the South Atlantic, which had eruptions about 2.5 million and about 40 000 years ago (Maund et al., 1988).

Massive outpourings of lava

In the north-western United States the Columbia River and Snake River Provinces are enormous effusions of lava. The two provinces are often treated as one, but are very different in age and geomorphology. Some cones are present, but probably most of the lava was erupted from long fissures. The Snake Province is essentially Quaternary and covers an area of 50 000 km², and the Columbia Province is of Miocene age and covers an area of 130 000 km². These massive outpourings filled valleys the size of the Grand Canyon, and totally changed the landscape.

The Deccan Traps of India cover an even greater area. The flows were erupted over a period of a few million years around the Cretaceous–Tertiary boundary. No original landforms are preserved, but the basalts are the original rock mass from which the landscape is carved (Figure 10.2).

From a geomorphic standpoint the interesting thing is that some of the older volcanic provinces may be related to global tectonics. The creation of new continental margins by the splitting of Pangea is likely to be accompanied by huge outpourings of lava. Many of these are confined to the sea-floor, but

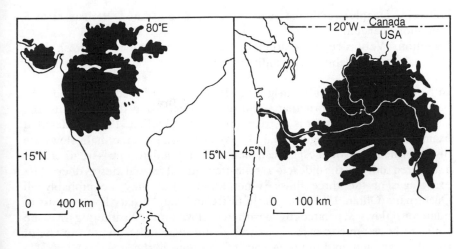

Figure 10.2 Extensive lava plateaux.
left: Deccan Plateau, India. Erupted about the Cretaceous–Tertiary boundary.
right: Columbia River Plateau, USA. The lavas are mainly of Miocene age.
(Source: C.D. Ollier, *Volcanoes*, Blackwell, 1988)

some such as the Deccan, Snake River and Karroo (South Africa) volcanics are preserved on land. The Columbia River Plateau may be related to a spreading site which has run from the Pacific under the North American continent. On all these volcanic provinces, geomorphic history begins not from some conjectural uplift or peneplain, but from new ground created by volcanic activity. The great thing is that volcanic rocks can be dated by both potassium argon dating and palaeomagnetism, so there is very good age control for the rocks and for landscape evolution.

Volcanoes and weathering

New lava flows are made of fresh rock, which eventually weathers, perhaps through a sequence of weathering products, over a period of time that may be measured in years, or millions of years. Western Victoria, Australia, includes an area of volcanic plains, mostly formed over the past few million years. It is possible to relate all the flows to individual points of eruption, marked by volcanic cones, so this is an example of 'areal' or 'polyorifice' eruption. The different flows can be distinguished by degree of weathering, and successive drainage modifications, together with absolute dating (mostly K/Ar dates on the basalts) enable a good evolutionary sequence to be worked out. Several terrains have been determined (Ollier and Joyce, 1987), as follows:

Eccles. Flows up to 20 000 years old. Virtually unweathered.
Rouse. Flows weathered, but with stony rises still evident (original flow surface features). Some ferricrete. Perhaps up to 2 million years old.

Dunkeld. Deeply weathered, with occasional corestones. Perhaps 2 to 5
  million years old.
Hamilton. Very deeply weathered, with no corestones or fresh rock
  remaining. Probably over 5 million years old.

In New South Wales palaeomagnetic studies of weathered basalt and sub-
basaltic soils have shown that some of the weathering dates back to the
Cretaceous although most is Tertiary (Schmidt and Ollier, 1988). A detailed
example is shown in Figure 10.3. In Western Samoa three main flows are
distinguished: the youngest is fresh rock; the intermediate one is weathered to
a stony red soil, and the oldest to a stone-free red soil several metres deep. The
exact ages of the three flows is not known, but they are probably all
Quaternary (Ollier, 1988). In Northern Ireland the Tertiary lavas consist of
numerous flows separated by weathered flow tops. Weathering may be
sufficiently extreme to make bauxites, giving some indication of the rate of
bauxite formation, and the frequency of eruptions in the area.

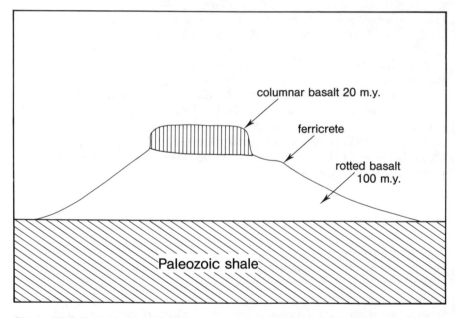

Figure 10.3 Cross-section of Bald Knobs, near Armidale, Australia. The upper basalt is
very fresh with splendid columnar structure, and a potassium/argon date of 20 million. This
rests on a ferricrete, which is at the top of a thoroughly weathered earlier basalt.
Palaeomagnetic readings on the weathered basalt gives an age of about 100 million years.

Inversion of relief

Liquid lava naturally tends to flow down valleys, and usually comes to rest
between opposing hillsides. The river that used to flow down the valley is

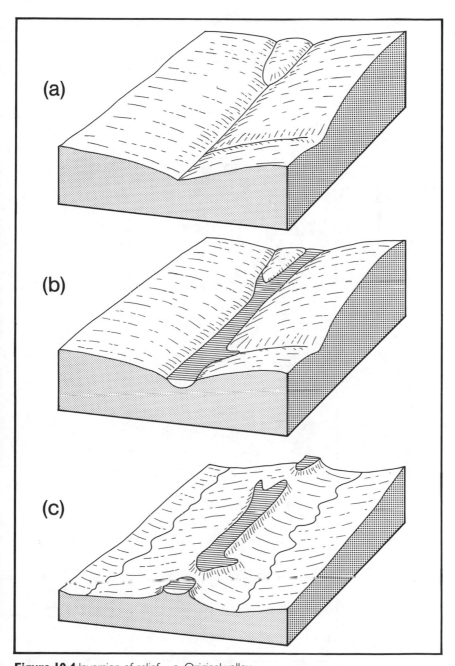

Figure 10.4 Inversion of relief.   a. Original valley
b. A lava flow fills the valley, and new streams (twin lateral streams) form on each side of the lava flow.
c. Downcutting and valley widening by the twin lateral streams leave the old valley-fill basalt on the ridge between valleys. (Source: C.D. Ollier, *Volcanoes*, Blackwell, 1988)

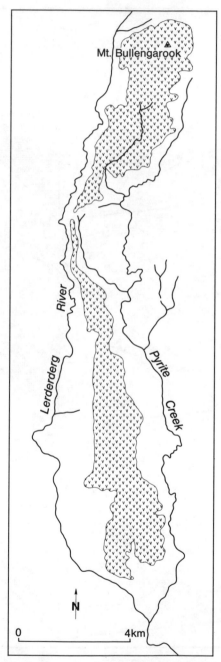

Figure 10.5 The Bullengarook lava flow, near Bacchus Marsh, west of Melbourne. The flat-topped lava flow is on a long ridge, and is bounded by the Lerderderg River and Pyrite Creek, which have cut down about 60 m below the top of the flow. The lava flow is about 3.5 million years old.

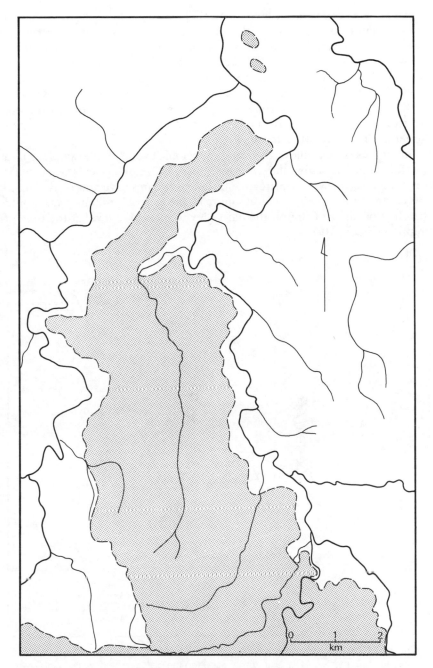

Figure 10.6 Inversion of relief at the 19 million-year-old Paddy's Plain flow from Ebor Volcano, New South Wales (see Fig. 4.4 for overall relationship).

displaced, and commonly comes to flow along the side of the lava flow (a lateral stream), or along both sides of the lava flow (twin lateral streams). Erosion of twin lateral streams creates a valley on each side of the lava flow. These may eventually cut down lower than the original valley, and the basalt (or other rock) of the lava flow is located on a ridge top. This is called inversion of relief (Figure 10.4). It may occur on plains dissected by occasional valleys, or on the flanks of volcanic cones. There are many stages in the creation of inverted relief, which can give a relative age-scale for such features. Examples are shown in Figures 10.5 and 10.6.

Sometimes lateral valleys give clues to tectonic events. For example, a lava flow filled the Aberfeldy valley in eastern Victoria 27 million years ago, when it was incised 420 m below a plateau. The lateral streams have now incised up to 330 m since that time. Tectonic uplift in the past 27 million years would therefore be almost 650 m if the old valley were at base level at the time of eruption (Figure 10.7).

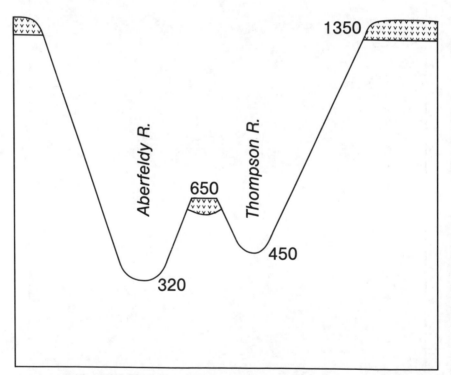

Figure 10.7 Cross-section of the valleys of the Aberfeldy and Thompson Rivers, Victoria, with elevations in metres. After eruption of basalt on the plateau, the river first cut down to the level of the lower basalt. Then, 27 million years ago, a lava flow covered the valley bottom, lateral streams developed and cut down to the present river level. If the Oligocene valley was at base level, then there was 650 m of post-Oligocene uplift. (Source: C.D. Ollier).

Drainage disruption

A volcano may be big enough to completely disrupt pre-existing drainage. Mount Etna is a fine example (Figure 10.8). The pre-volcanic drainage was to the south-east, but this was blocked by the volcano and drainage found its way around the north (R. Alcantara) and the south (R. Simento). Later lava flows have even diverted these displaced rivers to a lesser extent, as described by Chester and Duncan (1982). Mount Etna was erupted entirely in the Quaternary, and younger eruptions and diversions have occurred to the present time (Romano, 1982).

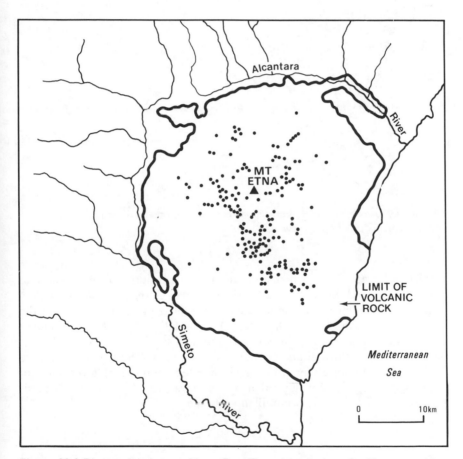

Figure 10.8 Disrupted drainage at Mount Etna. The volcano is about 2 million years old. Pre-volcanic drainage flowed NW–SE. Now the drainage is intercepted and flows around the volcano as the Alcantara and Simento Rivers. The many black dots are parasitic cones. (Source: C.D. Ollier).

A similar disruption is shown by the drainage of the Warrumbungle Volcano, New South Wales (Figure 10.9). This has clearly blocked the course

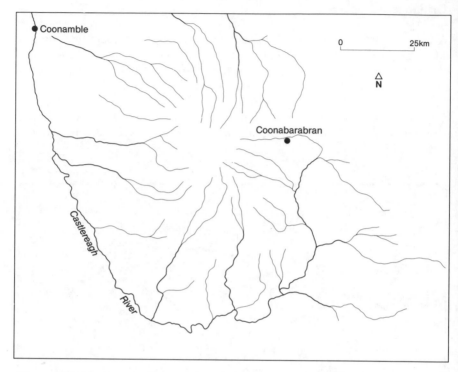

Figure 10.9 Diverted drainage of the Castlereagh River around the 13–17 million-year-old Warrumbungles Volcano, New South Wales.

of the Castlereagh River, which flows around the southern side of the volcano. This volcano, and the associated Castlereagh river course, is about 15 million years old. A much more complicated and older story is indicated by the Monaro Volcano in New South Wales (Ollier and Taylor, 1988). It is thought that drainage approaching the volcano from the south was once continuous with the palaeo-Murrumbidgee, but was diverted to the Snowy River (Figure 10.10). The Wannon River in Victoria is diverted not by a single volcano, but by a coalescing series of flows that make a lava plain (Figure 10.11). This diversion occurred about two million years ago.

Great Escarpments and volcanoes

Great Escarpments are large escarpments that run for hundreds of kilometres (see p. 119), separating a plateau from rugged country or lowlands. Volcanoes on the plateau that are intersected by the escarpment must be older than the

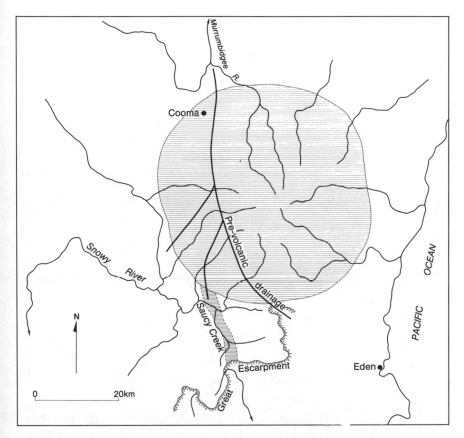

Figure 10.10 Diversion of formerly north-flowing drainage to the Snowy River by the 60 million-year-old Monaro Volcano, New South Wales. (Re-drawn after Ollier and Taylor, *BMR J Geol. Geophys.* 10. 357-62, 1988).

escarpment. An example is the Ebor Volcano, New South Wales (Figure 10.12). This volcano is about 19 million years old, so the escarpment here is younger than that. The 20 million-year-old Tweed Volcano lies on the coastal plain below the escarpment, so here the escarpment had already retreated past the site of the volcano by 20 million years ago. A lava flow in Queensland near Innisfail was erupted on the plateau and flowed over the escarpment (Figure 8.11). Rivers were formed on each side of the lava flow (North and South Johnston Rivers) and by now have cut sizeable valleys, but the 3 million-year-old flow is still essentially intact. The escarpment was therefore in existence by 3 million years ago, but it may have been present for very much longer.

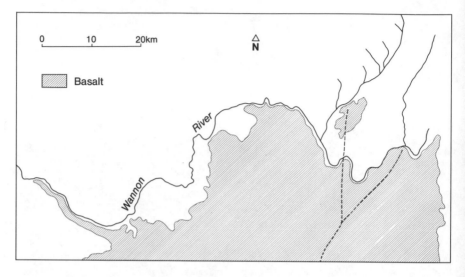

Figure 10.11 Diversion of the Wannon River, Victoria, around a lava plain built up about 2 million years ago.

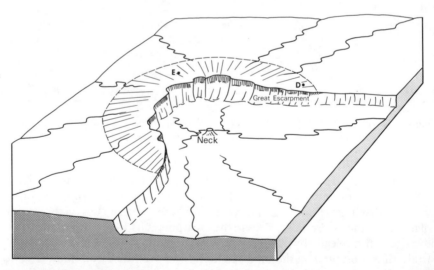

Figure 10.12 Block diagram of the 19 million-year-old Ebor Volcano, New South Wales, largely removed by the retreat of the Great Escarpment. (Based on Ollier, *J. geol. Soc. Aust.* 29, 431–35, 1982).

Faults, monoclines and lava flows

Faulting and volcanicity are often associated, and can be used to decipher local landscape history. A lava flow that has erupted after formation of a fault scarp is likely to flow down an individual valley (like the flow on Figure 8.11), and will pile up at the lower level. A lava flow erupted on a plain before faulting will simply be faulted, and will have the same thickness above and below the fault. A lava plain warped by a younger monocline, such as the Rowsley monocline west of Melbourne, will have an even thickness of basalt on all parts of the slope. In the Geelong area, Victoria, long lava flows occupied late Tertiary valleys, and were later warped by monoclinal folding. Since the original valleys could not flow uphill, the amount of warping can be worked out in detail, as well as subsequent modifications to the drainage pattern and river evolution.

Marine erosion

Oceanic volcanoes may build up a volcanic cone above sea-level, and initially this is likely to be a fairly simple cone. Plenty of examples are known, of which Surtsey is perhaps the most famous. Such a cone is immediately prone to attack from marine erosion, wearing it back and forming a cliff around the island and an offshore shelf. Tristan da Cunha is a fine example. The degree of cliffing may be a rough indication of age: Tristan, which erupted between 200 000 years ago and the present, is almost surrounded by cliffs 600 m high (Figure 10.13), but it is not nearly so eroded as its neighbouring volcanoes of Nightingale and Inaccessible which are a few million years old (Maund, 1988).

Figure 10.13 Tristan da Cunha Volcano in the South Atlantic. The cliffs are about 600 m high. (Photo: C.D. Ollier)

Figure 10.14 Ball's Pyramid, Lord Howe Island, New South Wales. The island is 551 m high with a base of 1.1 x 0.4 km, and is a remnant of a former volcano at least 6 km across. The island consists of a pile of almost level lava flows on a base of older lavas intruded by many dykes. (Photo: C.D. Ollier)

Eventually a volcano may be almost entirely consumed, and ultimately only a few fragments survive. Most often these are volcanic plugs, but occasionally the last remnant may be a simple succession of flows, like Ball's Pyramid, near Lord Howe Island (Figure 10.14).

Migration of volcanic centres

In Germany the spatial and temporal differentiation of Tertiary crustal movement is reflected in the age and distribution of volcanic events. The first eruptions took place in the Eifel in the Eocene, and continued into the Oligocene. During this period the centre of volcanicity migrated eastward. In the Miocene it moved to the Hessian upland and the Vogelsberg, a huge and important volcano at the northern end of the Rhine Graben. However, the easternmost volcanic area, the Rhone, was first active in the Oligocene.

The large Central Type Volcanoes of eastern Australia have a clear relationship of age with latitude (Figure 10.15). This is explained by Wellman and McDougall (1982) as the result of the passage of Australia over a hot-spot, as explained in Chapter 9. The age ranged from over 20 million years in the north to about 6 million in the south. It should be noted that there are many other volcanic eruptions in eastern Australia that do not fit this picture, so

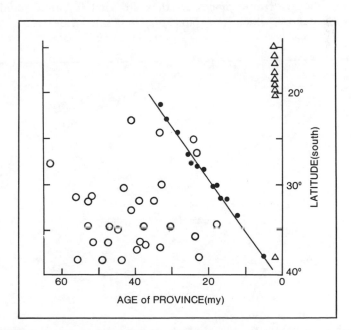

Figure 10.15 A plot of age against latitude of volcanoes of eastern Australia, after Wellman and McDougall (1982). The open circles are lava field eruptions: triangles are areal volcanoes including many maars and scoria cones: black dots are central volcanoes, and show a fine linear relationship thought to indicate passage of Australia over a hot-spot.

although some volcanoes are related to hot-spots, others seem to have quite different relationships (and incidentally different volcanic geomorphology). Seamounts in the Tasman Sea also appear to be related to hot-spots, and show the same rate of plate migration (p. 134.)

Volcanic arcs

Island arcs are a major topic in tectonic geomorphology, and many arcs are volcanic. There are many theories about their origin and relationships to plate tectonics or other features. There is no doubt that arcs with volcanoes of about the same age are formed today and were formed in the past. Different hypotheses require arcs to migrate in specified ways, but the actual distributions are inconsistent. The distribution of volcanoes of Papua New Guinea (Figure 10.16) shows two main arcs, as well as a scatter of volcanoes with no clear pattern. The northern arc has an inner Palaeogene arc and an outer (on the convex side) Quaternary arc, whereas the New Britain arc has the Quaternary volcanoes on the inside and the Palaeogene arc on the outside.

The arcs recognized by Locardi (1985) in Italy appear to show movement back and forth over the past 1.7 million years, with certain times being especially active (Figure 10.17). These volcanic ages are valuable not only for theorizing about tectonic processes: they are vital for understanding the geomorphic history, which is of course on a long time-scale.

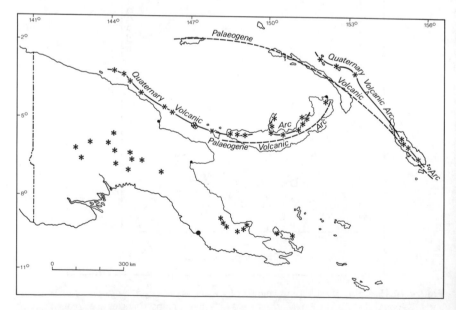

Figure 10.16 Distribution of volcanoes in Papua New Guinea. (Source: Ollier and Pain, Z. Geomorph. Supp. 69, 1-16, 1988)

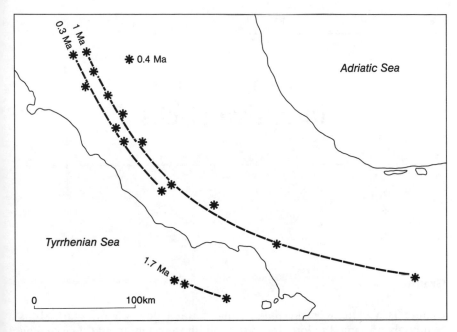

Figure 10.17 In Italy, central volcanoes with similar age of initial eruption are distributed along arcs that are related to extensional features. The arc migrates, but not systematically. (After Locardi, 1985).

Chapter 11
Intrusive rocks

Granite

Granite is a plutonic rock, that is, one formed deep in the earth. It is commonly presumed to form by crystallization from a melt, and slow cooling produces coarse crystals easily visible to the naked eye. Some granites appear to be an end result of metamorphism, that is, alteration of pre-existing rocks. The process of 'granitization' converts older rocks to the mineral composition of granite, but 'ghost stratigraphy' and other features reveal the metamorphic origin.

However it is formed, granite tends to be lighter than surrounding rocks and rises as plutons towards the earth's surface. Yet it cools as it rises, and granite is generally assumed to form at depths of several kilometres. Such granites can only be exposed at the ground surface after erosion of several kilometres of overlying rock, which would take a long time. Granites are younger than the rocks they intrude, and exposed granites are older than any unmetamorphosed rocks that overlie them. The age of granite can be determined by techniques such as potassium/argon dating.

In most areas where granite is exposed it is simply a basement rock, exposed by erosion and with limited direct relevance to landforms. In general true granites, rather rich in potassium, tend to be resistant to erosion and stand up as hills: less potassic, biotite-rich equivalents such as granodiorite tend to weather readily and give rise to lowlands, commonly surrounded by higher ground on the surrounding metamorphic rock (a metamorphic aureole, commonly consisting of the rock hornfels). The two types of granite, weathered and erodible or higly resistant, can occur in close proximity.

Individual plutons may rise from a more massive or wall-like batholith at depth, and as plutons rise they may spread into mushroom shapes, a phenomenon also noted in salt domes. Several geological techniques enable petrologists to determine to some degree whether a granite is of high or low-level origin. Many granites are now known to be tabular bodies, not extending to unspecified depths (Lynn et al., 1981; Hamilton and Myers 1967).

Products of plutonic intrusion, frequently a forerunner of uplift and mountain building, and its denudational products, such as conglomerates and fanglomerates, may give us some hints of such past events (Gansser, 1983). A striking example is the Bergell intrusion and its erosion products in the Como molasse (a coarse post-uplift sedimentary rock). The Bergell pluton was emplaced 30 million years ago into a mountain belt with nappes. Ganser says, without giving the evidence, that it must have crystallized under an overburden of at least 8 km. The overburden was rapidly removed during rapid uplift. Erosion products were deposited in the kilimetre-thick Como molasse, which is underlain by Middle Oligocene silts and overlain by Upper Oligocene sandstones and clays. Burgell boulders in the Como molasse are identical in age and petrology with those in the Bergell pluton. The conclusion is that there was rapid uplift of over 10 000 m of the Bergell, followed by erosion, transport and deposition in the Como molasse in a few million years. The evidence is supported by fission track studies which suggest the granite boulders in the molasse come from a position of 6 km above the present Bergell massif.

The Trans-Himalayan plutons have a similar overburden to that suggested for the Bergell. The biggest and fastest uplift of this range followed the intrusions of about 40 million years ago, producing a relief about 6000 m higher than that of today. The detritus is now found in associated molasse sediments, the lowest conglomerates containing granite boulders have Eocene clasts, so uplift and erosion was almost immediate.

The early history of the exposure of a granite may be determined by examination of strata in surrounding areas, where debris from material eroded from the granite is deposited. The Permian Mount Duval granite in New South Wales was already being eroded in the Permian, providing granite boulders for a .nearby conglomerate. Off the coast of southern New South Wales the Cretaceous and younger sediments of the offshore sedimentary wedge are especially thick opposite the Bega Granite (Davies, 1975), suggesting that it has been eroding at least since Cretaceous times.

Isotopic studies can help in relating granite instrusion to geological studies. The Willow Springs diorite in Death Valley has a U–Pb zircon age of about 11.5 million years. A 40Ar-39 analysis of hornblende gave an age of about 10 million years, showing that the diorite remained above 50g C for about 1.5 million years. Geological relationships show the granite was exposed by 5 million years ago (Asmeron et al., 1990). The rapid uplift and exposure occurred in an extensional environmental, as did the rise of the granites of Papua New Guinea (Ollier and Pain, 1980).

The Mole Granite in New South Wales was apparently emplaced very close to the surface, with no more than half a kilometre of overlying rocks (Kleeman, 1984). These have been stripped off, leaving the top of the granite as a structural plateau. The stripping probably occurred quite early, and the Mole

Plateau landform has probably looked very much as it does today since the Triassic. The granite intrusion itself is of interest, being a tabular intrusion (Figure 11.1). Calculations suggest it has a maximum thickness of 4 km, and a more probable thickness of only 1 km.

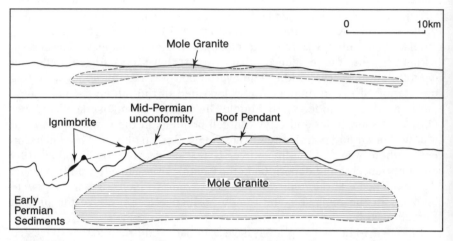

Figure 11.1 Cross-section of the Mole Granite, New South Wales. Top is natural scale, bottom is 10 x vertical exaggeration. the intrusion is tabular, and the location of the feeder pipe is not known. This granite was intruded very close to the ground surface, as indicated by the roof pendant, the Mid-Permian unconformity and the Ignimbrite. (After Kleeman, 1984.)

A somewhat similar situation was described from northern New South Wales by Korsch (1982) at Mount Duval. It seems that a large volcano was erupted on to a palaeoplain, over a granite magma chamber, and then the granite continued to rise until it intruded its own volcano. In doing so the granite reached a height of some hundred metres higher that the palaeoplain. The ascending granite gave rise to a contemporaneous synclinal rim depression around the granite in which coarse sediment accumulated which later became a congolmerate. The conglomerate contains fragments of the Duval granite, indicating very rapid exposure of the granite. Mount Duval now rises above a surrounding palaeoplain, as it probably did very soon after its emplacement. The scenery looked very much the same in Permian times as it does today.

Many granites are deeply weathered, and the age of the weathering can sometimes be determined. Of course weathering is still going on today, and in some tropical areas present-day weathering is sufficiently active to prevent exposure of fresh rock except in actively eroded river beds, as on the 1.9 million-year-old Fergusson Island granite in Papua New Guinea. In this situation it is rather pointless to ask when weathering started. Elsewhere the question makes sense. The Bega Granite in southern New South Wales was already very deeply weathered in Paleocene times when it was partly covered

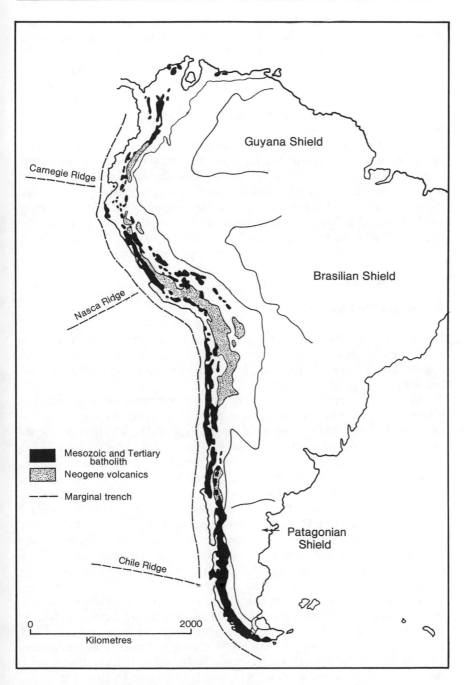

Figure 11.2 The distribution of Mesozoic and Tertiary batholiths and young volcanics in the Andes. (Source: C.D. Ollier, *Tectonics and Landforms,* Longman, 1981)

by the Monaro Volcano. Erosion of the weathered granite provided huge amounts of sand deposited on the continental shelf since Cretaceous times.

Some granite topography, as distinct from the age of the granite rock, can be dated in relation to neighbouring sediments. Mount Buffalo in Victoria is a granite plateau dissected by valleys which have been partly filled by sediments. Most valleys contain Tertiary gravels, but some contain Permian conglomerates. The topography at Mount Buffalo was very similar to that of today in Permian times.

Emplacement of granite is often associated with uplift of mountains, though the age of the granite and highlands may be very different. A long line of granite batholiths runs along the Andes Mountains (Figure 11.2). A similar line of Palaeozoic batholiths underlies the eastern highlands of Australia, though the uplift did not occur until the Mesozoic or even Tertiary in some parts. A line of granites, some as young as 1.4 million years, underlies the mountain backbone of New Guinea.

Gneiss mantled domes

Domes of the metamorphic rock gneiss with a core of granite are well known in ancient rocks, such as the Precambrian of Scandinavia. Much younger examples are known in Arizona and in Papua New Guinea which have a direct bearing on topography.

The New Guinea gneiss mantled domes are at the eastern end of a belt of granite that underlies the mountain backbone of the country, and was presumably the cause of its uplift (Ollier and Pain, 1981). The best developed dome is the Goodenough Island Dome (Figure 11.3). This is about 12 km across and 2 km high. It is bounded by triangular facets rather like the planeze remnants on volcanoes, but cut in metamorphic rocks. The foliation of the gneiss is parallel to the surface of the dome, and concentric with its outcrop. The foliation developed at the same time as the intrusion and the formation of the topographic dome. The dome is cored with granite which has been dated at 1.9 million years. Numerous other domes occur in the area. Other plutons rise from the long batholith that underlies the central mountains of New Guinea, including the Ok Tedi granite which has been dated at 1.4 million years.

Tectonic denudation is the name given to a process whereby crust is stretched and a new land surface is essentially created by tectonics. Holm et al. (1990) describe tectonic denudation in the Neogene in the Black Mountains in Death Valley, California. They suggest that an original zone of crust perhaps as little as 10 km wide was stretched to 150 km during uplift along a major detachment zone. The Black Mountain block is one of the youngest, and perhaps deepest exposed examples of Cordilleran metamorphic core complexes. Other examples are described in the section on detachment faults in Chapter 13.

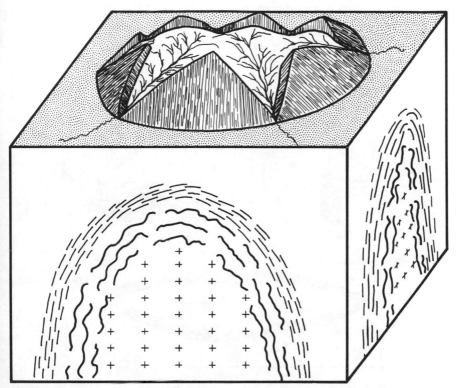

Figure 11.3 Block diagram of the gneiss-mantled dome of Goodenough Island, Papua New Guinea (after Ollier and Pain, 1981). The granite is rising to above the general ground surface, and originally had a cover of metamorphic rock with foliation parallel to the dome surface.

Intrusive volcanic rocks

The lavas that erupt in volcanoes originate deep in the earth's crust and reach the surface through fissures and pipes. When erosion has removed upper portions of volcanoes and neighbouring bedrock, the intrusive igneous bodies give rise to landforms, which may be used to date whole landscapes. The age of the volcanic rock can often be determined, but the landforms will be much younger as much time may be spent in removing their cover. The commonest kinds of volcanic intrusion are shown in Figure 11.4

Volcanic necks

Volcanic necks are the cylindrical feeders of volcanoes which are full of solidified volcanic rock. They usually erode at a different rate from their surroundings, commonly more slowly to form steep hills, but occasionally faster, giving rise to basins.

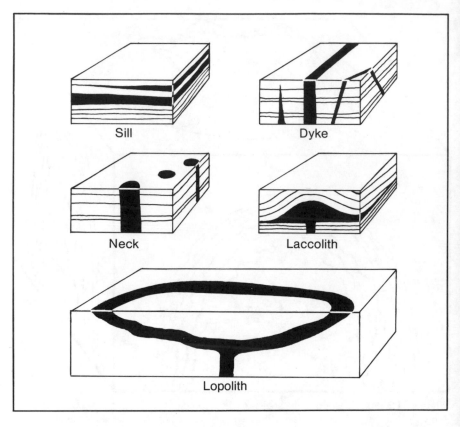

Figure 11.4 Common types of volcanic intrusion. (Source: after C.D. Ollier, *Volcanoes*, Blackwell, 1988).

Dykes

Most volcanoes are fed through vertical fissures which give rise to vertical sheets of igneous rock called dykes. They are usually a few metres wide, and commonly occur in swarms. Dykes give rise to wall-like ridges, such as the Breadknife at the Warrumbungle Volcano, New South Wales (Figure 11.5), but sometimes give rise to a trench-like depression.

Dykes fill cracks, or in other words the crust moves apart to make space for them. If there are many dykes, the crust must have expanded many times, and the total width of dykes gives a measure of the amount of crustal extension. In conjunction with dating, this can give a measure of a very important feature of the earth's surface — the rate of its growth, and at the same time the date of crustal extension. The Tertiary dyke swarm in Mull, Scotland, reaches an aggregate thickness of 1000 m, and shows a stretching of the crust in the affected area of 3.8 per cent. The dyke swarm in Arran indicates a stretching of 7 per cent.

Figure 11.5 The Breadknife, a dyke in the 13-17 million year old Warrumbungles Volcano, New South Wales. (Photo: C.D. Ollier).

The sheeted intrusive complex of the Troodos Range in Cyprus consists of a swarm of N–S basic dykes invading a series of older lavas. The dykes make up 90 per cent of the whole, and being nearly vertical, their total thickness shows that their intrusion involved E–W crustal extension of over 130 km in Palaeozoic times. This has no direct impact on present landforms, of course, and is mentioned here only to illustrate the method. Using the same method, it is clear that Iceland, being crossed by many dyke swarms, must be larger today than it was in the past. The rate of extension, measured from the width of dykes emplaced since the Pliocene, is 5 cm/year, which can be taken as the present rate of sea-floor spreading in the North Atlantic. At that rate the Atlantic would take only 100 million years to form.

Sills

Intrusive rocks may form nearly horizontal tabular bodies of rock called sills, especially when intruded into horizontal strata. They can have very large dimension, like the Jurassic sills of Tasmania, which are up to 700 m thick. Intrusion of sills 'floats' off the overlying rock, which must have a considerable effect on the topography. Sills also cover large areas: the Karroo sills in South Africa extend over an area of 500 000 km². Other kinds of volcanic intrusion are described in textbooks of volcanology (e.g. Ollier, 1988a).

Mantle plumes and intrusive doming

Plumes are related to hot-spots described in Chapter 8 and 9. The plume hypothesis postulates that plumes of hot mantle material rise as relatively narrow columns that spread laterally at a high level to produce areas of anomalously hot asthenosphere up to 2000 km across. The dynamic and thermal effects may produce up to 2 km of uplift in the centre of a broad dome. Fracturing of the dome may produce rifts and massive outpourings of basaltic lava. This may be the start of a triple junction, and may even go on to the stage of sea-floor spreading.

Cox uses the Plume–Dome model to explain features of the Deccan of India (intruded about the Cretaceous/Tertiary boundary), southern Brazil (where the Atlantic opened about the Lower Cretaceous), the Karoo of south-eastern Africa (which separated from Antarctica in the lower Jurassic). Cox bases his interpretation on drainage pattern mainly, which has a crude fit. The examples are all those used by geomorphologists in the study of Marginal Swells (*Randschwellen*) and Great Escarpments, and is a variation on the theme of intrusion or thermal causation of uplift at continental margins.

Major volcanoes are usually associated with an underlying intrusive complex, in the form of plugs, plutons, cone-sheets, ring dykes and linear

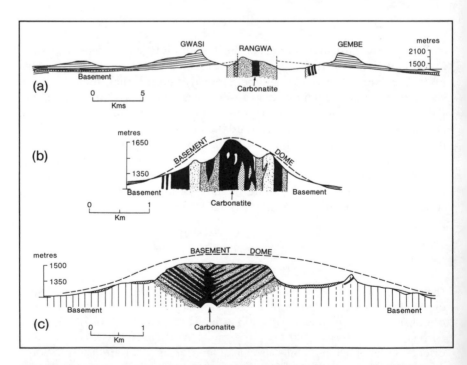

Figure 11.6 Examples of doming up of the basement beneath volcanoes of East Africa. a. Kisingiri. b. South Runi Hills. c. Homa Mountain. (After King, Le Bas and Sutherland, 1972.)

dykes. These intrusive rocks push up the ground surface, so the volcano appears to have errupted onto a pre-existing hill. The total uplift may be more than half a kilometre. Examples include Napak in Uganda, Kisigiri in Kenya, Koolau Volcano in Hawaii, and Tweed Volcano in Australia. Even if the volcano is entirely eroded, the attitude of the surrounding rock may still indicate the basement dome (Figure 11.6). The intrusive complex causes a positive gravity anomaly, a magnetic anomaly, and has a volume about one-third of the volume of volcanic rock (Wellman, 1986). The intrusive rocks present no problem, but it would be easy to misinterpret the nature of the pre-volcanic geomorphology if the nature of these intrusions was not understood.

Chapter 12
Faults and rifts

Faults

A fault is a fracture through rock along which there has been displacement of the sides relative to one another parallel to the fracture. There are several

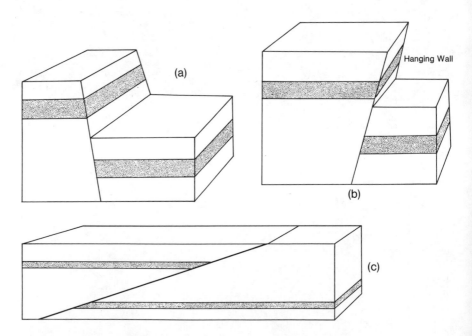

Figure 12.1 Some common faults. a. normal fault (tensional). b. High-angle reverse fault (compressional). c. thrust fault (compressional). (Source: C.D. Ollier, *Tectonics and Landforms*, Longman 1981)

kinds, of which the simplest are normal faults, caused by tension and extension, reverse of thrust faults, caused by compression and shortening (Figure 12.1). Strike-slip faults (also known as transverse or transcurrent faults) are nearly vertical faults along which the opposite sides have moved in opposite horizontal directions parallel to the fault plane.

Faults are seen in the landscape as ridges or escarpments along the fault. If the topography is a direct result of the fault displacing the ground surface, the scarp is called a *fault scarp*. If the topography merely results from differential erosion on opposite sides of an old fault it is called a *fault-line scarp*. The arrangement of faults gives several recurring types of structure which have been given names such as graben, horst, and tilt block (Figure 12.2). The nature of faulting is described in many textbooks of structural geology, and geomorphology textbooks usually explain the main landscape aspects of faults. Here we will concentrate on the age of faults and associated landscapes.

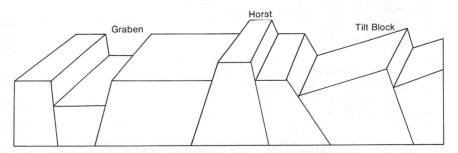

Figure 12.2 Diagrammatic representation of some common fault-generated landforms. (Source: C.D. Ollier, *Tectonics and Landforms*, Longman, 1981).

Normal faults, horsts, grabens and tilt blocks

A fault must be younger than the rocks that are faulted. An example of successive faults is shown in Figure 12.3, and the dated lava flows enable a detailed interpretation of the landscape evolution over the past 13 milllion years.

Faults may be put into a relative sequence in relation to drainage and other features in the vicinity. The Lake George Fault, for instance, cuts across drainage lines, which must therefore be older than the fault (Figure 12.4). The old Taylor's Creek used to flow right across the area from east to west, but is now dammed back to make a lake in the fault angle depression. To the north a number of creeks that formerly flowed north-west are diverted by the fault, and after barbed junctions they flow along the base of the fault scarp into Lake George. To the south of Lake George the fault scarp runs across the Molonglo River. This has a much bigger catchment than the northern rivers and was able to maintain its course across the rising fault block, an example of antecedent

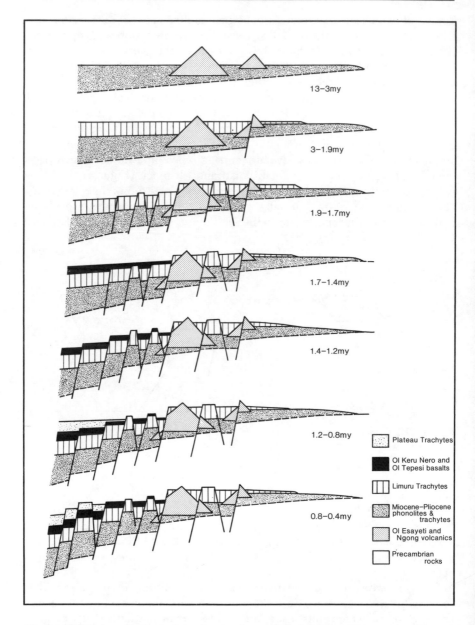

13–3my

3–1.9my

1.9–1.7my

1.7–1.4my

1.4–1.2my

1.2–0.8my

0.8–0.4my

Plateau Trachytes

Ol Keru Nero and
Ol Tepesi basalts

Limuru Trachytes

Miocene–Pliocene
phonolites &
trachytes

Ol Esayeti and
Ngong volcanics

Precambrian
rocks

Figure 12.3 Daigrammatic sections showing the evolution of the eastern margin of the south Kenya rift. (After Baker and Mitchell, 1976.)

drainage. It is unusual to find three different responses to faulting — diversion, damming and antecedence — in such a small area.

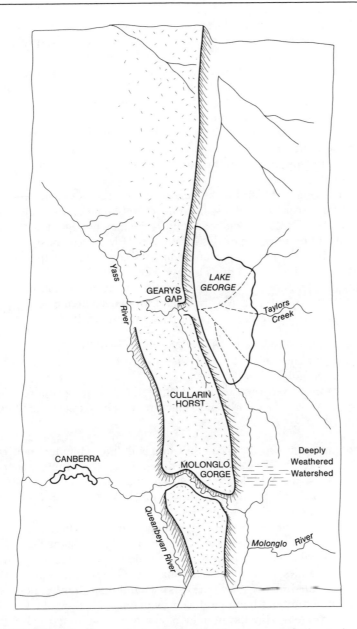

Figure 12.4 The Lake George Fault and associated features, near Canberra, Australia. For interpretation see text.

The geomorphic interpretation does not stop there. The old rivers were flowing across a planation surface cut across Palaeozoic rocks, so the surface must be post-Palaeozoic and pre-Miocene — in fact it is probably of Palaeozoic/Mesozoic age. The surface was in some places deeply weathered.

Most of the weathered rock has been removed on the upland, but is preserved on the more protected downthrow side. The weathering is probably of Mesozoic age, and has in turn had an effect on the Early Tertiary sediments such as those found in Geary's Gap, the abandoned channel of the old Taylor's Creek at the top of the fault scarp. Thus the fault and its associated features enables us to work out the age of many other landscapes features, back to Mesozoic times or possibly even earlier.

As time goes on fault scarps are eroded, and the longer they have been in existence the more eroded they will be. Other factors such as climate and lithology are also important, and this method is not much use in actually dating escarpments.

Horsts are elongated fault blocks, relatively uplifted between two normal faults. At a small scale the Cullarin horst near Canberra is an example (Figure 12.4). At the large scale a whole mountain range may be regarded as a horst. The central and eastern High Atlas Mountains of Morocco consists of rocks folded in the Jurassic, but the present mountains are a range shaped as a unit by the formation of a horst in the Tertiary (Schaer, 1987).

Faulting may be periodic, or occur at specific times in geomorphic history. In the Yellowstone Park area, Wyoming and Montana, three main periods of evolution have been determined (Barnovsky and Labar, 1989):

1. Erosion dominated between 18 and 17 million years ago.
2. Block faulting and subsidence of basins occurred between 16 and 14.8 million years ago.
3. After 14.8 million years ago and before 8.6 million years ago uplift occurred in northern Yellowstone Park, and Yellowstone Valley strata were tilted and eroded.

Movement may occur repeatedly on the same fault line, a phenomenon called resurgent tectonics. Wallace et al. (1990) describe faults in Montana that have been active from the Middle Proterozoic to the Holocene — perhaps 700 million years. At least some of the faulting would have direct consequences on landscape evolution.

Normal faulting and extension can also occur in areas that are otherwise considered to be compressional, such as island arcs. The magmatic arc of the Antarctic Peninsula, for example, displays a wide range of features related to extensional tectonics at a convergent margin (Garret and Storey, 1987). Subduction finished about 50 million years ago, and since then a series of horst and graben structures has been produced by extension.

In the convergent Himalayan region extension faulting is also found (Royden and Burchfiel, 1987). East–west striking, gently north-dipping faults in the high Himalayas and southern Tibet, at least 600 km long, formed during the post-collisional convergence of India and Tibet. They are thought to be of Late Miocene age, and confined to the upper crustal levels.

A fault angle depression may form at the foot of a fault scarp, especially if the downfaulted block is back-tilted. Sediments may accumulate in a fault angle depression, and the age of the oldest sediment gives the minimum age of the fault. Thus at Lake George near Canberra the oldest sediments are probably Miocene, so the fault is at least of Miocene age.

Another example of a fault angle depression is the Ovens Valley area in Victoria. This originated in the Permian, and contains Permian glacial deposits as well as younger sediments, as described on p.24. Although the Permian faulting did not directly produce the landforms of today, it would not be possible to account satisfactorily for all the geomorphic features of the area without knowing of the Permian precursors of the present landscape. Some large fault angle depressions are known as half-grabens. These will be described later in the section on rift valleys.

The Basin and Range Province

A special case is provided by the Basin and Range Province of the United States, which can be regarded as an area of tilt blocks. Cainozoic extension has doubled the width of the Basin and Range Province. Faults reach depths of 20 km and range in age from Oligocene to Holocene (Hamilton, 1987). The traditional view of the Basin and Range Province, that it is formed by high-angle normal faults is slowly changing, for more modern work shows there is a great deal of low-angle normal-slip faults that have accommodated profound crustal stretching (Davis, 1984). Many of these detachment faults are of Oligocene–Miocene age. They are cut and offset by the conventional Basin and Range high-angle normal faults. Detachment faults have also been called low-angle normal faults (LANF), denudational faults, decollements, and lag faults. Wernicke (1981) described low-angle normal faults in the Basin and Range Province

Another example comes from Death Valley, California, where the Panamint Range is a structural block detached in the Cainozoic from rocks that now form the Black Mountain and transported 80 km to the north-west (Stewart, 1983). Transport took place along a single detachment fault, and is dated by volcanic rocks associated with the chaotic blocks on the upper block at 6 to 8 million years.

In recent years attention has been focused on metamorphic core complexes of the Basin and Range Province. These highly extended areas are characterized by an upper plate of generally unmetamorphosed to slightly matamorphosed rocks, and a lower plate of matamorphic and plutonic rocks. Separating these two dramatically different structural styles are sub-horizontal detachment surfaces.

Most components in an extending system must rise towards the surface with time (Hamilton, 1987). Most detachment fault literature is about Mid-Tertiary complexes, but Hamilton (1987) describes Late Neogene detachments in Death Valley. Here are the highest steep scarps, the greatest deformation of Quaternary materials, the deepest topographic closures of structural depressions and the youngest exposed detachment faults of Late-Miocene to Holocene age. An example is the Tucki Mountain detachment fault in California. The upper plate consists of unmetamorphosed Palaeozoic strata in the east, Pliocene slide megabreccias on the crest and Pliocene sediments and slide breccias on the west. Hamilton believes his model may also apply to Eocene domes in NE Washington and British Columbia, to Palaeogene to Miocene domes in the island Naxos in the Aegean Sea, and to Mesozoic to Mid-Tertiary domes in the Betic Cordillera of south-east Spain.

Listric faults

Many extensional faults are curved in section, getting flatter with depth. Such faults are called listric faults. An example is the Rio Grande rift shown in figure 12.5, The Rio Grande rift is about 850 km long, running roughly north–south from Colorado to New Mexico. There has been about 8 km of crustal extension near Albuquerque since about 27 million years ago (Woodward, 1977). The rate of spreading is 300 B, which can be compared with the rate of 400 to 1000 B for the eastern rift system of Africa (Baker, Mohr and Williams, 1972,), and 100 B for the Rhine graben (Illies, 1972).

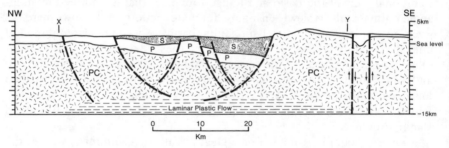

Figure 12.5 Diagrammatic section across the Rio Grande rift near Alberquerque, New Mexico. Extension of 8 km is calculated between points x and y. S sedimentary graben fill; P Palaeozoic to Oligocene rocks; PC Precambrian rocks. (After Woodward, 1977.)

Another example of listric faults is shown in Figure 12.6, a morphotectonic sketch of the Tyrrhenian Sea. The basic topography consists of fault blocks, but the listric faults only affect the upper part of the crust, and at greater depths plastic deformation probably occurs. In the centre of the basin, beneath the Bathyal Plain, the crust is thinned. The faulting occurred in Pliocene and Quarternary times (Wezel, 1982b).

Low-angle thrusts

Low-angle thrust faults are still active in some places, but it is difficult to relate them directly to landforms because local deformation is rather complex. Elter and Trevisian (1973) showed that the average velocity of 50 km-wide nappe was about 10 000 B, and roughly similar measurements have been made elsewhere. Most thrusts are basically inactive and exposed by erosion. Being

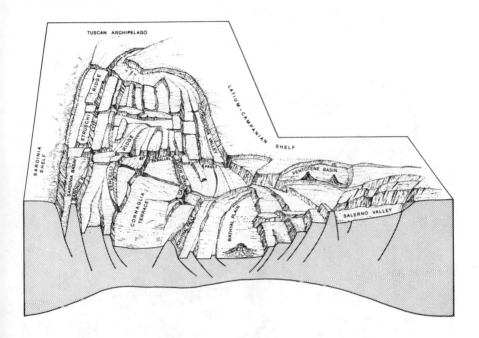

Figure 12.6 Sketch (not to scale) of the Tyrrhenian Sea. Note the listric faults and the thinned crust beneath the Bathyal Plain as revealed by seismic survey. (After Wezel, 1982b.)

low-angle features they have sinuous outcrops, with 'windows' exposing material beneath the fault plane, and 'klippen' where patches of material above the thrust are isolated as outliers beyond the main outcrop of the thrust. The klippen landform — a hill — must be younger than the thrust fault, as must be the valleys around it.

To the east of the northern Rocky Mountains lies an area known as the 'Disturbed Zone', where Mesozoic and early Tertiary rocks of the Great Plains are overturned and thrust by easterly movements of the mountains. Low-angle thrust sheets such as the Lewis Thrust have moved eastward several tens of kilometres. Precambrian rocks come to lie over Cretaceous and younger rock. Heart Mountain is an outlier of the Lewis Thrust and is a classic example of an erosional klippe. The thrusts are demonstrably gravitational glides (Reeves, 1946; Pierce, 1957). They have evidently slid over the existing relief at the ground surface.

The margin of the middle Rockies also has strongly folded Palaeozoic and Mesozoic rocks thrust from the west over Tertiary formations. The southern Rockies have Precambrian nuclei that were upwarped during the Palaeozoic and Mesozoic, and then affected by further vertical movement in the Laramide diastrophism, usually dated about 65 million years ago. Eardley (1963) pointed out that the largest uplifts droop over surrounding basins and so are gravitational, not compressional.

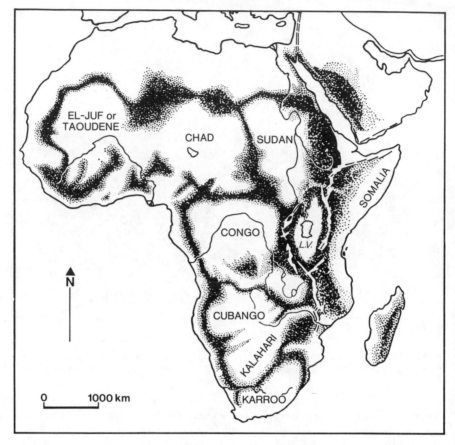

Figure 12.7 The basins and swells of Africa, and the associated rift valleys. (After Holmes, 1965.)

Rift valleys

The continent of Africa is divided into a number of huge basins by swells (Figure 12.7), and rift valleys follow the lines of some of the swells. Rift valleys are huge grabens, elongated downfaulted areas, hundreds of kilometres long. The African ones have a total length equal to one-sixth of the world's circumference. Rift valleys have remarkably uniform widths, as shown below (width in km):

Albert Rift	35–45
Tanzania	40–50
Nyasa	40–60
Dead Sea	35
Rudolf	55
Baikal	55–70
Rhine	30–45

The constancy of width is remarkable, since many rifts are known to be growing wider at rates of about 100 to 1000 B.

They are bounded by normal faults with huge throws, and the base of the graben may be below sea-level. The basement of the Dead Sea is as low as −2 600 m; that of the Nyasa Rift is −1005 m. Opposite sides of the rift may be at different heights. They are often, but not always, associated with a rise towards the rift which makes a divide some distance from the actual fault.

The East African rift system can be dated by associated volcanism. In the western rift volcanism began about 12 million years ago in the north and finished at 7 million years ago in the south, the volcanism being either before or concurrent with initial development of the rift sedimentary basins. The general picture is similar in the Kenya rift valley, where volcanicity started earlier but still migrated to the south. In eastern Kenya rifting commenced in the Jurassic.

The Gregory Rift of Kenya is largely an asymmetrical half-graben on the Kenya Dome (Hackman et al., 1990). Further north in Kenya the Turkana Rift consists of a series of half-grabens of alternate polarity, with a volcano in the middle of the half-graben (Dunkelman et al., 1988). The rifts, which connect the Kenya Rift to the Ethiopian Rift, were active from the late Cretaceous to the Pleistocene, and cut across clear domes.

The Ethiopian Rift Valley has been described in part by Woldegabriel et al. (1990). The central sector had six major volcanic episodes between the late Oligocene and the Quaternary, which helps to date the physiographic evolution. A thinned Mesozoic stratigraphic sequence along the rift margin suggests that doming may have preceded volcanism and rifting of the Cainozoic. By late Miocene (10 to 8 million years ago) the east and west faulted margins had formed. There was a two-stage rift development:

1. In the Oligocene to Early Miocene alternating half-grabens were formed.
2. In the Late Miocene and Pliocene symmetrical rifts were formed, with a neovolcanic zone in the middle.

The Munesa Crystal Tuff (3.5 million years old), a prominent marker on both margins and revealed at depth in a drill hole in the rift shows 2 km of downthrow in the central sector since its eruption. The breakup of Arabia from Africa (Figure 12.8) started at 40 to 50 million years ago in the East African Rift and concentrated from about 20 million years ago around the Red Sea (Courtillot et al., 1987). The sea-floor spreading that formed the Gulf of Aden began in the Oligocene and most of the uplift of the plateau margins had been achieved by the end of the Oligocene. Further sea-floor spreading took place in the Late Miocene (Abbate et al., 1986).

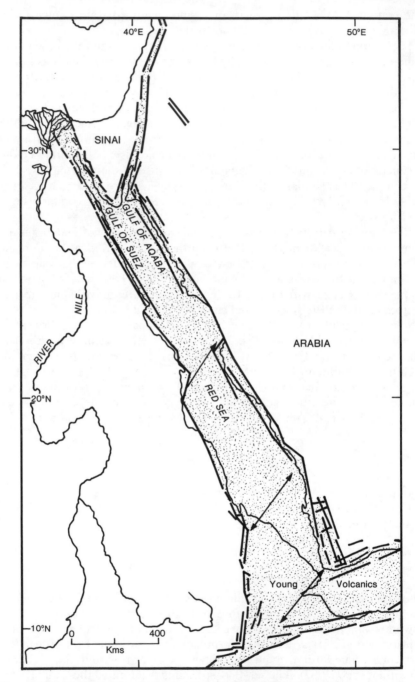

Figure 12.8 The separation of Arabia from Africa by the opening of the Red Sea rhombochasm and associated features. (Source: C.D. Ollier, *Tectonics and Landforms*, Longman, 1981).

The Suez Rift is a structure 300 km long and 50 to 80 km wide. The opening started 23 million years ago (Miocene). According to Chenet et al. (1987) the opening was followed by uplift of the shoulders and increasing subsidence of the central trough. The structures of the present-day rift result almost entirely from the extension that has been active since rift initiation, and no significant deformation occurred earlier.

The subsidence of the Rhine Rift Valley began in the middle Eocene (Bremer, 1989). It was especially intensive in the south in the early Tertiary, and in the north it was active in the Upper Tertiary and continues to the present time. There are traces of marine influences in the south dating from the Oligocene and in the north from the Miocene.

Lake Baikal in Siberia is commonly regarded as a rift valley, though in some ways it is better regarded as a half-graben. It is of enormous dimensions and holds most of the world's fresh water. The rift follows a huge fault on the western side for over 450 km bounding the Siberian Platform. It is very asymmetrical, but there are faults on both sides. The basin was mainly formed by rapid subsidence during the past 3–4 million years (Artyushkov et al., 1990), but the sediments in the depression show that the rift zone had its origin in Oligocene/Miocene time, and the large domes which are morphotectonic precursors go back to the Late Mesozoic (Ufimtzev, 1990). It is bounded by normal faults which indicate extension of the crust during subsidence of about 3–7 per cent. In the Early and Middle Miocene, smooth hills and wide shallow valleys existed in the Baikal area. At its southern end some minor valleys are antecedent, but the lake overflows via the River Lena. There appears to be an old erosion surface that pre-dates the faulting; its age is not clear, but is possibly Mesozoic.

The development of the Baikal rift zone is thought to mark an initial stage in the break up of the Eurasian plate. Intrusion of an asthenosphere intrusion reached the base of the crust about 30 million years ago, according to Zorin (1981). The Baikal rift grabens do not form a single valley along the crest of the Boikal arched uplift, but form a complicated branching system of depressions. Continental crust underlies the depression, but it is 7–10 km thinner than the crust under the adjacent mountains. Zorin concludes that formation of the rift occurred as a result of plastic extension (necking). In the case of Lake Baikal extension is about 25 km, but in associated depressions (Tunka, Chara, etc.) it is just a few kilometres. Extension was irregular both in space and time. The Baikal rift had two main stages of development: a first stage began in the Oligocene and lasted to the Middle Miocene. Downwarping was slow and relief generation rather low. The second stage began in the Upper Pliocene (2–3 million years ago), and is still in progress.

Another kind of rifting in a totally different environment has been described from the Caribbean (Mann and Burke, 1984). Here there are sixteen Cainozoic rifts in an extension zone on the northern edge of the Caribbean Plate. They formed successively by localized extension. Some are on land today, in Jamaica, Central America, Honduras and Guatemala. The oldest rifting occurred in the Paleocene or early Eocene, and the youngest is still active. There is a continuity of morphotectonic activity from the early Tertiary to the present.

Strike-slip faults

Horizontal movement on opposite sides of a fault is called strike-slip faulting. Transform faults are a special variety of strike-slip fault found on the sea-floor and associated with sea-floor spreading that just occasionally extend under continents, and the term should not be applied loosely. Strike-slip faults are often very long (hundreds of kilometres), and frequently active over a very long period.

Perhaps the world's best known fault, the San Andreas Fault (Figure 12.9), is a major fault zone that relates to sea-floor spreading and transforms faults in the Gulf of California. The total length is over 1 000 km and a displacement rate of about 3–5 cm/year has been determined by geodetic surveying since 1970. To move 1000 km at this rate would take only about 25 million years, but nevertheless the major landscape features associated with the fault must be viewed on a long time-scale. Two sedimentary basins that were contiguous in the Early Miocene have been separated by 320–330 km of right-lateral displacement (Stanley, 1987). Faulting occurred during the Early Miocene and later. This gives an average rate of movement of only about 16 mm/year. The Alpine Fault of New Zealand is another famous strike-slip fault, and its similarity to the San Andreas Fault is made clear in Figure 12.9

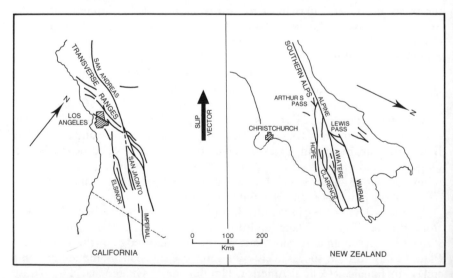

Figure 12.9 Fault maps of southern California and South Island, New Zealand, rotated so that the slip vectors are parallel for the two regions. (After Scholz, 1977.)

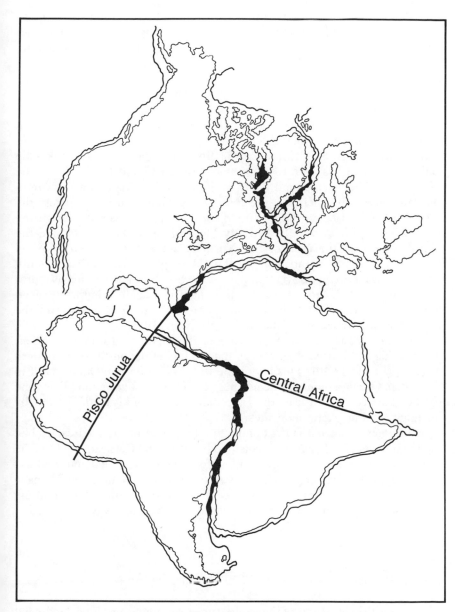

Figure 12.10 The Pisco Jurua fault as a continuation of the North Atlantic Rift (after Szatmari, 1983); and the Central African Lineament as a continuation of the north coast of South America (after Cornacchia and Dars, 1983). Some faults develop into new oceans, others have relatively minor geomorphic effects.

The Dead Sea is a site of strike-slip faulting and crustal spreading, with the creation of new sea-floor. The sideways movement is neatly demonstrated by the absence of a delta at the mouths of the Wadi Zarqua and the Wadi Mujil, and yet the presence of a delta with no apparent source of sediments in the El Lisan peninsula (Figure 8.3). If the delta was built by these wadis in former times, they have since been separated by about 40 km. Using this and other evidence Quennel (1959) found that lateral movement in the region amounted to 107 km, of which 62 was from Early Miocene to Early Pliocene, and the second phase of 45 km was from Middle Pleistocene to the present. The landscape history must therefore be traced back to the Early Miocene, say 25 million years ago. Similarly the Red Sea–Gulf of Aden region started to split about 25 million years ago.

In north-western South America the Bocono Fault is a major strike-slip fault. Movement has been estimated at between 15 and 100 km since the Mesozoic and between 69 and 100 m in the Holocene. Another major strike-slip fault in South America is the Pisco–Jurua fault, which cuts across the continent from Pisco on the Pacific to the Guyana–Surinam border on the Atlantic coast (Figure 12.10). In Early Mesozoic time the fault formed the south-western continuation of the North Atlantic Rift. Its northern end opened as the North Atlantic Rift opened. The separation of North and South America caused the north-western part of the South America to move south-west along the Pisco-Jurua fault, creating the Tacutu graben in the Guyanan Shield, the gently folded Jurua zone across the Amazon Basin, and the Pisco–Abancay deflection of the Andes (Szatmari, 1983).

A rather similar fault or fault zone is the Central African Lineament (Cornacchia and Dars, 1983), continuing the line of the north coast of South America (Figure 12.10). This fault zone runs W–E from the Guinea Gulf to the Indian Ocean, bordering the northern rim of the Congo craton. This zone has been active since at least the Late Precambrian and is marked today by seismicity and magnetic anomalies.

Sometimes strike-slip faults dominate a large part of the landscape. The topographic relief of Panama is explained by Mann and Corrigan (1990) as the result of north-west movement and deformation accommodated across a diffuse zone by two major sets of strike-slip faults, both of which have prominent physiographic expression. The plate collision and faulting responsible for the morphotectonics occurred in the latest Miocene to Early Pliocene.

Chasmic faults

At one time the continents of the world were associated as one huge landmass, Pangaea. This broke up, and the creation of new sea-floor separated the continents into roughly the positions they are in today. It seems to be accepted that a rift valley stage precedes the wider separation of continents, and traces of rift valleys may be found around some continents. However the major faults that extend right through the lithosphere are different in kind from simple normal faults, and the name 'chasmic fault' has been suggested for them (Osmaston, 1973). Today they mark the edge of the continental masses, and are offshore. Nevertheless they do relate to geomorphic features, in determining the nature of the offshore zone and most importantly, its age.

As soon as a continental margin is isolated and bounding the sea, the continent will start to erode and deposit sediments along the new continental margin. The Atlantic opened in the Jurassic, so no pre-Jurassic sediments can be expected on the sea-floor, but Jurassic and younger sediments will be found around the continental margins.

In Australia the different edges of the continent were formed at different times. The north-west possibly faced the ocean since Jurassic times, the Tasman Sea between Australia and New Zealand opened between about 80 and 60 million years ago, and the separation of Australia and Antartic started about 55 million years ago. This should result in different kinds of geomorphology around the continental edge, and different deposits on the continental shelf. This is only partly true, and one of the tasks for geomorphology in Australia is to work out why.

Chapter 13
Mountains

Just about the most obvious landforms on earth, mountains, are scientifically a difficult topic. This difficulty comes not from the mountains themselves, but because of the questions scientists ask about them, and the assumptions they make. A few decades ago mountains were explained in terms of geosyncline theory. Books about mountains tended to be full of theory about geosynclines, but little on the topography. Indeed, when Hall (1859) first suggested the geosynclinal theory to account for geology of the Appalachians, he was accused by one of his critics of having a theory of mountain-making that left the mountains out.

Today plate tectonics is the fashion, and many books about mountains are mainly about plate tectonics and have very little on geomorphology or even elementary topographic description of mountains. You might expect that a book entitled *The Anatomy of Mountain Ranges* (Schaer and Rodgers, 1987) would describe the form of mountains as well as their insides and speculation about their former physiology, but in fact almost all is speculation about structure, and could just as well have been done on structures of the plains of the world wherever they are amenable to a plate tectonic reconstruction. All the chapters in *Mountain Building Processes* (Hsu, 1982) are the same except for one by Gansser, who emphasized the 'morphogenetic phase' which makes mountains what they are — high land — and dealt with the processes that make mountains, not the rock structures inside them.

It has long been observed that some rocks are deformed, or in simple terms 'folded.' The idea arose that some force in the earth squashed the rocks into folds, and at the same time the squashed rocks became mountains. The search for this force became the search for the mechanism of mountain-building. One idea was the squashing of sediments in a sedimentary trough or geosyncline, as the bounding masses of hard rock, the cratons, came together. Another idea was that sediments at the edge of a continent (North America for instance) were folded and pushed into mountains as the continent drifted: in North America forming the Rockies and Coast Ranges, in South America the Andes.

The ruling theory in tectonics today is plate tectonics, in which folding and mountain-building are related to subduction sites, where one slab of earth's crust is thrust beneath another, causing folding and uplift. This is essentially a variation on the same fold-and-uplift scheme as the older ideas. A variation on the plate tectonics theme is that an underthrust slab, perhaps with subducted sediments, is carried well under the overriding slab where it melts into granite, which rises and forms mountains some distance, perhaps hundreds of kilometres, from the trench that presumably marks the subduction site.

Significant features of mountains include the following:

1. Mountains are not confined to folded rocks.
2. Mountains are not confined to collision sites.
3. Folding is not confined to collision sites
4. Folding is not confined to mountains: many lowlands are developed on folded rocks.
5. Folding can be caused by many mechanisms and may never have been associated with mountains.

In reality the forces that cause folding and those that build mountains are quite different, but even professional geologists confuse them.

Terminology

There has long been a tendency to confuse the folding of rock with mountain-building, exacerbated by the advent of plate tectonics. The confusion is clearly expressed in the definitions of Bates and Jackson (1987):

orogeny. Literally, the process of formation of mountains. The term came into use in the middle of the 19th Century, when the process was thought to include both the deformation of rocks within the mountains, and the creation of the mountainous topography. Only much later was it realised that the two processes were mostly not closely related, either in origin or time. Today, most geologists regard the formation of mountainous topography as postorogenic. By present geological usage, orogeny is the process by which structures within fold-belt mountainous areas were formed, including thrusting, folding, and faulting in the outer and higher layers, and plastic folding, metamorphism, and plutonism in the inner and deeper layers. Only in the very youngest, late Cenozoic mountains is there any evident causal relationship between rock structure and surface landscape.

This modern usage has displaced the usage of Gilbert (1890) who wrote: 'The process of mountain formation is orogeny, the process of continent formation is epeirogeny, and the two collectively are diastrophism.' Epeirogeny is still a valid term, but has also shifted in meaning. According to Bates and Jackson (1987):

epeirogeny. As defined by Gilbert (1890), a form of diastrophism that has produced the larger features of the continents and oceans, for example plateaus and basins, in contrast to the more localised process of *orogeny,* which has produced mountain chains.

The last part of the statement does not seem to match their own definition of orogeny. To understand anything about mountains, including their age, it is necessary to be very careful of the terms orogeny, epeirogeny, mountain-building and folding, and be aware that different authors use these terms in different ways, and a single author may have assumptions or shifts of meaning in usage of these terms. *In brief, there is no necessary relationship between folding and mountain-building.*

Mountain-building

In reality mountains are not made by folding, but by erosion of uplifted land, which may or may not consist of folded rock. The first important process is uplift. Uplift of a former lowland makes a plateau, which may still exist, and dissection of a plateau makes more rugged 'mountains.'

The question 'What is the age of mountains?' must be broken up into several separate questions, such as the following:

What is the age of the rocks of which the mountains are formed?
What is the age of folding of the rocks (if any) in the mountains?
What is the age of planation of the mountain region, if it were ever planated?
When was the area uplifted?
When was the erosive phase initiated?

The point of these abstract questions may become clearer after a few examples have been given.

Most writers on mountains are not content to ask these 'when' questions, but go straight to the much harder question, 'How were the mountains uplifted?' Many different mechanisms of uplift to make plateaux and mountains have been proposed, and some of these relate to the age of the landforms. A list of twenty possibilities is given in Table 13.1. In this book I want to concentrate on age, and not on mechanisms of mountain uplift.

Classification of mountains

Mountains may be classified in many ways. Howell (1989) for instance classifies them into:

Tensional
Compressional
Transcurrent
Thermal

This is essentially a classification of presumed modes of origin of mountains, not of the mountains as landforms.

For present purposes it is desirable to have a geomorphic classification, based on what can be seen rather than speculated on, though after this theory can play a part. Mountains are classified into the following categories:

A. Dissected Plateaux
B. Tilt Blocks
C. Border Mountains and Median Plateaux
D. Marginal swells and Great Escarpments
E. Alpine Mountains
F. Exotic Terrane Mountains
G. Domes

A. Dissected plateaux

Perhaps the simplest plateaux are simple horsts or uplifted fault blocks. A series of north-south-oriented in Western Sichuan and Yunnan Province of south-west China have been given the name of Hengduan Mountains, which literally means the cross-cutting mountains. The area has complex geological

Table 13.1 Some possible causes of tectonic uplift

1. Thermal expansion due to a mantle plume or hot-spot.
2. Thermal expansion due to overriding and subduction of a hot mid-ocean ridge or spreading centre.
3. Thermal expansion due to shear heating along a lithosphere–asthenosphere interface.
4. Expansion accompanying partial melting (the increase in volume on fusing basalt is about 8 per cent).
5. Hydraulic reactions such as serpentinization (10 per cent expansion).
6. Introduction of volatiles due to deep-seated dehydration of hydrous minerals.
7. Expansion due to depletion of 'fertile' mantle in garnet and iron resulting from basalt genesis.
8. Crustal thickening due to horizontal transfer of mass in the lower crust.
9. Deep-seated solid state reactions such as eclogite-basalt transformation.
10. Subduction at a very shallow angle, perhaps horizontal.
11. Simple subduction, or continent/continent subduction.
12. Cessation of subduction and resulting thermal equilibrium of static slab.
13. Isolation of a plateau by listric normal faulting in the surrounding area.
14. A piece of cooling lithosphere detaches from the crust and is replaced by a counterflow of asthenosphere, which warms the crustal rocks and causes uplift (due to thermal expansion) and volcanism.
15. Intrusion of magma into the lower crust.
16. Intrusion of sills.
17. Isostatic uplift after scarp retreat.
18. Isostatic rebound after regional erosion.
19. Underplating, the addition of unspecified lighter material to the base of the crust.
20. Isochemical phase change with volume increase in the lower crust or upper mantle.

structure, but was planated into an extensive erosion surface in the Tertiary, and is essentially made by Quaternary uplift of long, narrow horsts and graben. Differential uplift has resulted in dissected plateau remnants at many different levels, ranging from about 1700 m to 4700 m (Liu and Zhong, 1987).

The Appalachians

The Appalachians consist of a very thick sequence of Palaeozoic sedimentary rock. The rocks have been folded and faulted, also in the Palaeozoic and especially the Late Palaeozoic. By the Jurassic many structures were planed off, so the area must have been above sea-level by that time (though not necessarily mountainous) to allow erosional planation. Meyerhoff and Olmsted (1963) believe that the present drainage lines are direct descendants of Permian streams, so planation may have started as early as the Permian. The area was possibly covered by Cretaceous sea, but was certainly reduced to very low relief in Cretaceous times.

Broad general uplift raised the subdued surface into a plateau, which still exists as such features as the Allegheny Plateau. It is important to note that although the geophysical data of the Appalachians is usually interpreted in terms of Palaeozoic collision tectonics, there was also a phase of Mesozoic extension which is usually overlooked or downplayed (Heck, 1989). This is especially significant if the geomorphic history also starts in the Mesozoic. After uplift, erosion of the plateau created the present-day Appalachian Mountains. The whole area has superimposed drainage and is related to the continental margin and offshore deposition. Since the North Atlantic Ocean came into existence by sea-floor spreading in the Jurassic, the tectonic history has an approximate fit with the geomorphic history.

Sheldon Judson (1975) wrote that in Jurassic time the drainage of the Appalachian region was to the north-west on to the north American craton. The north-westward regional tilt was reversed to the south-east as the continental margin subsided as a result of the divergence of the North American plate from the spreading centre now seen as the mid-Atlantic ridge. Beginning with that reversal the modern streamways and associated topography have developed. Old sinkholes, containing fossils of Paleocene and even Cretaceous times, have been found on planation surfaces hundreds of metres below the present summits (Rodgers, 1983; Bridge, 1950; Pierce, 1965) so the uplift history must go back to at least then.

The Rocky Mountains

The Rocky Mountains are rather far inland to be explained by plate tectonic subduction. The mountains are in reality a dissected plateau. Atwood (1940) wrote:

Many persons who have studied the Front Range of Colorado have called the remnants of this old-age erosion surface in that section the Rocky Mountains Peneplain. A remnant in the San Juan Mountains in southwestern Colorado has been called the San Juan Peneplain.

He lists many other plateaux in the area: Green Ridge Peneplain, Medicine Bow Peneplain, etc.

The edge of the Rockies are complicated by gravity-spreading and thrust-faulting as described in Chapter 12. Some of the thrusts moved over existing ground surfaces, perhaps around 60 million years ago. The geomorphic expression of these areas has not been described in detail.

The southern Rocky Mountains and their continuation into Mexico, according to Eaton (1987), are the crest of a symmetrical continental feature of large dimensions, which he calls the Alvardo Ridge. It is characterized by gentle rises which were originally covered by continental sediments of Miocene and Pliocene age, tens to hundreds of metres thick. On the eastern rise the blanket is well preserved and has been undisturbed for nearly 5 million years. In the Canadian Rockies, deformation occurred as a series of intense pulses spread over 100 million years from Late Jurassic to Paleocene. Deformation terminated with uplift of the foothills belt in the Paleocene.

Western British Columbia is a collage of exotic terranes. The present high relief of the Coast Mountains is a product of rapid late Cainozoic uplift (Muhs et al., 1987). Nevertheless, an erosion surface of moderate relief existed over most of southern British Columbia in Early to Middle Miocene time. The Miocene landscape was characterized by broad valleys, and lavas erupted on to this surface when it was near sea-level are found on summits up to 2.5 km elevation. Much of the elevation is therefore post-Miocene. In contrast, north of 52°N Miocene lavas or the erosion surface on which these lavas were deposited extend into valleys 1 to 1.5 km below mountain summits, suggesting that some of the present height of the central and northern Coast Mountains is pre-Miocene in age. Fission track dates confirm that the Coast Mountains were raised 2 to 3 km in the last 10 million years (Parrish, 1983).

The Andes

The Andes are often cited as the type example for plate tectonic explanation of mountain-building, with subduction of the Pacific Plate under the South American Plate creating the Andes by compression. The hypothesis is also supposed to account for the generation of granites under the Andes, eruption of volcanoes and of course folding of rocks in the Andes Mountains. In fact there is abundant evidence to show that much of the Andes was reduced to a plain in the Tertiary, and has since been uplifted as a series of fault-bounded plateaux, including the high plains or altiplano. Volcanoes were erupted on to these plains, but the story is not easy. There are regional variations and conflicting interpretations.

A simple version of the Andes in Peru is shown in Figure 13.1, in which the planation and main blocks were already in existence 100 million years ago. Since that time landscape evolution has continued, by tectonic uplift, volcanism and erosion. The story has been described in more detail in some parts of the Andes (e.g. by Mortimer, p.94), but the essential feature is that the Andes have not been made by folding, and that landscape evolution goes back about 100 millions years.

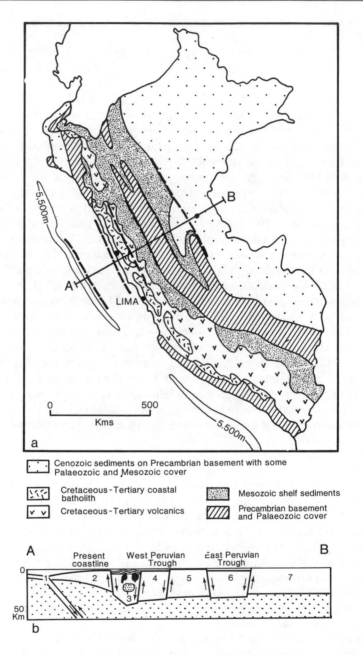

Figure 13.1 Top. Simplified geologic and tectonic map of Peru (After Myers, 1975). Bottom. Section AB of map above showing the structure of Peru 100 million years ago, and the relative movement of blocks. Subduction is indicated, but it seems to be irrelevant to the movement of blocks or the formation of mountains. 1. Oceanic crust; 2. Paracas Geanticline; 3 and 4. West Peruvian Trough (3. Paramonga Block; 4. Chavin Block); 5. Maranon Geanticline; 6. East Peruvian Trough; 7. Brazilian Shield. (After Myers, 1975.)

The end of the Cretaceous and the emplacement of the Andean batholith marked a distinct change in Andean history. In the north-west of Peru Tertiary marine clastics in fault basins, and red continental sediments were deposited in elongate basins parallel to the present trend of the Andes between the east and west cordilleras. 'Development of the modern Andes began in the late Tertiary. Normal faulting has been of major importance. Intermittent periods of standstill allowed the formation of several widely recognized erosion surfaces, some of which are now conspicuously tilted' (Jenks, 1975).

In Peru the Andes may be divided into three: eastern, western and a less distinct central. The Cordillera Occidentale is the most continuous, and consists of folded and faulted Palaeozoic and Mesozoic strata cut by Upper Cretaceous and Tertiary intrusives, and covered in the south by great thicknesses of volcanics. A pulse of Miocene volcanism, common throughout the Central Andes, is also present in the Cordillera Occidentale. Ashflows aged 8 to 11 million years fill deeply incised palaeovalleys, showing the Cordillera was uplifted before the Late Miocene. Several high, wide plains between the eastern and western cordillera, such as the Late Titicaca Basin, appear to be normal faulted tectonic basins largely filled with Tertiary and Quaternary sediments, and are not erosional plains.

In Colombia the Andean chain fans out into three: the western, central and eastern Cordillera. During the Cretaceous a great inundation invaded not only the central Cordillera but also a large part of the Guyana Shield. Early in the Tertiary the sea withdrew towards the present coastal area and the structural features became increasingly accentuated.

More details of the timing of uplift are available for particular regions. In the Colombian Andes, for instance, some areas were already uplifted above the (present) forest line as early as 16 million years ago. In the eastern Cordillera fission track analysis indicates a major uplift phase around 9–12 million years ago. Most other areas above 3000 m reached their present altitude only after 6 million years ago, by more recent uplifts or by the eruption of stratovolcanoes on dissected planation surfaces (Kroonenberg et al., 1990).

In Ecuador the Andes Chain has an eastern (Cordillera Real, the core of the Andes made of high grade metamorphic rocks) and western side (Cordillera Occidentale, consisting mainly of Cretaceous rocks), with the central Quito–Cuenca Depression between. This is flanked by Quaternary volcanoes, making the 'Avenue of Volcanoes' of Humboldt. The onset of uplift in the Eocene uplifted the Andes Ranges, which replaced the Guyana Shield as the principal source of sediment. There was further movement in the Miocene, and again the Pliocene and Quaternary.

In Chile from west to east there are three morphotectonic units: the Coastal Range, the Longitudinal Valley, and the High Cordillera. The latter gradually rises to the plateau of the Puna de Atacama at about 4000 m above which are volcanoes reaching to 6000 m. The southernmost part of the Southern Andes differs from the main Andean chain in being inactive and aseismic (Zeil, 1979). The area had late Cretaceous granite intrusion followed by rapid uplift and erosion according to Milnes (1987), but he fails to provide any details of geomorphology.

The Kimberley Plateau and Carr Boyd Ranges

The Kimberley Plateau in north-west Australia is the dissected remnant of a plateau cut across varied rocks. The Carr Boyd Range is an extension of the plateau on to more steeply dipping strata, the 'ranges' being bevelled strike ridges (p.81). The erosion surface was glaciated, and patches of till remain, covering striated pavements. The glaciation is the Sturt Glaciation of about 700 million years ago (Proterozoic), so the plateau already existed in Precambrian times. Stratigraphic evidence from surrounding basins suggests the area has never been covered by younger strata.

Antarctic Mountains

Tingey (1985) has described several Antarctic Ranges. The Prince Charles Mountains are made up of large flat-topped massifs near the main drainage glaciers and isolated outcrops away from the main ice streams, the whole having accordant summit levels, thought to be a pre-glacial erosion surface of low relief. Removal of rock by glacial erosion has resulted in about 1 km of isostatic uplift. The Transantarctic Mountains, one of the world's major mountain ranges, have been uplifted at least 3 km since Jurassic times. Uplift of the Transantarctic Mountains during the Cainozoic occurred despite glacial loading of the continent and the lack of any obvious tectonic cause. The mountains were apparently uplifted through a pre-existing ice sheet.

The idea that glaciation in Marie Byrd Land was accompanied by vertical uplift and volcanism was proposed by Le Masurier and Rex (1983). They emphasize the exceedingly flat pre-volcanic erosion surface which they think formed in Late Cretaceous/Early Tertiary times. Using that surface as a datum they infer vertical uplift of 3–5 km in the past 60–80 million years and possibly within the last 28 million years.

Greenland

Before the Tertiary central east Greenland was an extensive planation surface cut across Precambrian gneiss (Brooks, 1985). Continental breakup in the area was heralded by Late Cretaceous basin formation with deposition of sediments followed in the Palaeogene by voluminous basalt extrusion. Morphostructural features developed at this time include a major flexure parallel to the coast, a large dome with a diameter of 200 km and a height of 6 km, and regional plateau uplift of about 2.5 km. The first two occurred at about 50–55 million years ago, and the third at about 35 million years ago.

B. Tilt block mountains

Some mountains are the dissected remains of tilt blocks caused by faulting. It is normally assumed that the tilt block is a purely tectonic feature which later suffers erosion. However, Koons (1990) has suggested a feedback mechanism between erosion and uplift, at least if the tilt block is related to subduction. Subduction may originally create a tectonic wedge on the overriding slab, steeper on the subduction side and gentler on the side away from subduction. Erosion, partly controlled by orographic rain induced by the new mountains, will concentrate erosion near them, and especially at the foot of the steeper side. The rapid erosion will in turn lead to more rapid uplift by isostatic response, high heat flow, and exposure of deep crustal rock assemblages. The gently sloping side of the wedge is likely to have relatively low erosion rates, upper crustal material, and possibly the preservation of older landforms. Ruwenzori in the middle of the African plate could not be affected by subduction, but there could still be some erosional feedback once an initial tilt block has been produced.

The Sierra Nevada (United States)

The Sierra Nevada is the remains of a huge tilt block, with a massive fault separating it from Death Valley, 85 m below sea-level. Sedimentary strata were folded in the Late Palaeozoic and intruded by huge batholiths. This had been eroded to a plain by Tertiary times, and the uplift and tilting to the west happened in the Late Tertiary. The single colossal block of the Sierra Nevada was covered by volcanic flows, especially to the north.

Huber (1990) has provided more detail. A precursor of the Tuolumno River in the central Sierra Nevada cut a large valley that was filled by volcanic rocks 10 million years ago. This allows various calculations and reconstructions. The uplift of the crest of the range over the past 10 million years has been calculated at 1830 m, which may be compared with 2150 m estimated from the San Joaquin River 30 km to the south.

Ten million years ago an ancestral range of hills occupied the present site of the Sierran crest. Although it had only moderate relief it was a barrier to westward drainage, even before the Cainozoic uplift. At that time the San Joaquin was the only river flowing west across the range. The modern channel of the Tuolumno River is 915 m lower than the abandoned channel, and as much as 1525 m of channel incision has taken place in the last 10 million years (an average rate of about 150 B).

The Ruwenzori Mountains

Ruwenzori is a dissected massif of Precambrian rocks that has been uplifted to a great height, and makes by far the highest non-volcanic mountain mass in Africa. To the west lies the Western Rift Valley of Uganda, which started to accumulate sediment in the Oligocene or Lower Miocene, which approximately dates the commencement of uplift. The eastern side of

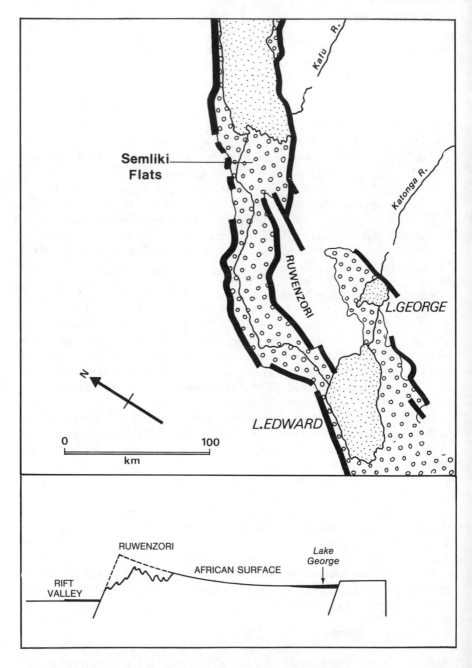

Figure 13.2 Top. Map of Ruwenzori in relation to the southern end of the Lake Albert Rift Valley.
Bottom. Cross-section across the Ruwenzori tilt block and Lake George. A section further north would show a horst, but this dies out to the south to become a simple tilt block.

Ruwenzori is simply warped and merges into the African erosion surface without a break (Figure 13.2). The age of the erosion surface in this part of Africa is poorly controlled, but is probably Palaeozoic–Mesozoic.

The Basin and Range Province (United States)

The Basin and Range Province is a classic example of fault block topography, and can be well dated by associated volcanoes. This area was discussed in Chapter 12.

In China the western part of Yunnan Province resembles the Basin and Range Province. It is formed by listric faults parallel to the eastern edge of the basin, which have produced steep fault scarps, dissected into triangular facets. Faulting was intiated in the Neogene, and there are up to 1900 m of Neogene sediments over Palaeogene sandstones and shales (Wu and Wang, 1988). The Sierra de Famatina in north-western Argentina appears to be a tilt block uplifted along a major thrust fault. It became a positive feature for the first time about 6 million years ago (Tabbutt, 1990).

C. Border Mountains and Median Plateaux

Zagros Mountains, Elburz Mountains and Iranian Plateau

Holmes (1965) illustrated the concept of median areas by the Iranian Plateau, bounded on one side by the Zagros Mountains, on the other by the Elburz Mountains. It is significant that although the Zagros Mountains and their thrust faults may be conveniently ascribed to subduction of the Arabian Plate, there is a mountain range on the other side of the plateau, with opposite direction of thrusting. Various tectonic mechanisms can be invoked to explain the situation. Holmes (1965, p.1112) suggests that the outward thrusting from the median area suggests either the approach of the forelands, or gravity sliding from the Iranian Plateau when it stood much higher, or both. From the landscape point of view the important feature is that the edges of a plateau are being uplifted to make mountains. This is, topographically, merely a variation on making mountains out of plateaux.

The thrust faults of the Iranian Plateau are of Tertiary age. It is interesting that the modern mountains on the ridge fit yet other schemes. Farhoudi (in press) describes the morphogenetic phase of the Alborz (Elburz) Mountains. The modern mountain range has a north-west trend, and post-dates a phase in which rocks of Palaeozoic and Mesozoic age with an easterly trend were exposed. A similar situation exists further north-west in the Caucasus Mountains. The folded rocks of the region trend north-west, but it is the Transcaucasus uplift that makes the modern mountains trend north.

The Zagros Mountains are a classic area for gravity tectonics, and many features of the complex geomorphology and drainage could be explained by structural response to the cutting of valleys. However the anomalous drainage can be explained in other ways (Oberlander, 1965)

The Himalayas, Kunlun Mountains and the Tibetan Plateau

The Tibetan Plateau is the highest plateau in the world, but is bounded on both its sides by even higher mountains. The Himalayas with south-verging thrusts face India; the Kunlun Mountains with north-verging thrusts face China. The rocks involved are mainly Mesozoic and Tertiary. The Tibetan Plateau has marine Cretaceous strata, gently folded, so uplift is post-Cretaceous. Some of the thrusts are still active, but they were also active through the Tertiary.

In the Himalayas there is also an isostatic response to the cutting of the major valleys through the plateau edge. The Himalayan valleys, kilometres deep, are so huge that they reduce the weight of the earth's crust so significantly that the crust rises, like a boat which has been unloaded. The erosion accounts for the edges rising higher than the original plateau: erosion of the valleys is actually causing the mountains to rise higher. Since rivers from the plateau cross the Himalayas in antecedent courses (p.32) they were in existence before the Himalayas grew to their present height. The most spectacular gorge, the Indus Gorge is about 6 km deep and only 21 km wide.

Complex evidence has been used to determine the timing of Himalayan uplift, including evidence derived from a study of the Bengal Fan, a huge submarine fan built of detritus derived from the Himalayas (Copeland and Hanson, 1990). The fan dates from the present to 18 million years ago, and is essentially first-cycle detritus derived directly from the uprising mountains. A substantial amount was deposited in Early to Mid-Mesozoic times, and is attributed to a pulse of uplift and erosion in the Himalayas. Nevertheless, some part of the Himalayas was suffering rapid erosion (1000 to 10 000 B) throughout the Neogene. The lack of granulite rocks in the eastern Himalayas suggests that uplift was distributed in brief pulses over different parts of the mountain belt, and is not compatible with models that suggest uniform uplift over 40 million years, or rapid uplift in the past 2 to 5 million years.

Further evidence comes from the study of fission track ages of detrital zircons from sandstones of the Siwalik Group — a sedimentary pile derived from the Himalayas. For at least the last 18 million years there have been areas in the Indus River catchment with uplift rates, relief and erosion rates, rates comparable to those observed in the Nanga Parbat region today (Cerveny et al., 1988). In effect the contemporary Himalayan landscape, on a broad scale, has been a relatively steady-state feature for at least 18 million years.

Gansser (1983), after considering the evidence of pluton intrusion and the age of molasse deposits associated with the Trans-Himalayan mountains concluded that the morphogenetic uplift of the Trans-Himalayas must have happened with various pulses, following successive intrusions.

D. Marginal swells and Great Escarpments

The popular concept of a mountain range is a tent-shaped symmetrical ridge with a sharp, angular top. But the name 'range' has often been given to features which are in reality escarpments — steep slopes on the side of a plateau. Some of these are known as Great Escarpments, having lengths of thousands of kilometres and heights of hundreds of metres. They may have a general name, or may be given many geographical names which tends to hide their unity.

The Great Escarpments are not simply scarps on plateau edges. They are restricted to continental margins, especially those margins which in plate tectonic terms are called passive margins. Some have structural control of horizontal rocks, but many do not.

Clearly an escarpment must be cut into some high ground, and around the continental margins there tends to be a swell, also parallel to the continental margin. These features have already been described in Chapter 8. In English we can call them marginal swells or marginal bulges. In German they are *Randschwellen,* and in French they are *bourrelets marginaux.* A range of marginal swells have been described by Godard (1982) and Ollier (1985).

Southern Africa

The Great Escarpment of southern Africa was reviewed by Ollier and Marker (1985). It runs in a large curve from Namibia to the Limpopo River (and no doubt further if a larger area had been studied), and the Kalahari Basin is bounded by the marginal swell behind the escarpment. The high erosion surface is probably of Mesozoic origin, and the escarpment started to retreat after continental breakup in the Jurassic. De Wit (1988) suggested that the Great Escarpment in the west of South Africa is related to a pre-Karroo period and not to the breakup of Gondwanaland, which would make it a very exceptional escarpment.

The Drakensberg in South Africa is essentially a Great Escarpment facing the Indian Ocean, and above it is plateau — the Highveld. The rocks are generally horizontal, and the upper part consists of 1500 m of Triassic basalt. Nowhere are there good constraints on its age. It is presumably associated with the opening of the surrounding oceans in the Jurassic and Cretaceous.

India

The Great Escarpment (Western Ghats) of peninsular Indian was described by Ollier and Powar (1985). The creation of the escarpment was related to the formation of a new continental margin with the breakup of Gondwanaland and the opening of the Arabian Sea, which here occurred at the Cretaceous–Tertiary boundary, the time of extrusion of the Deccan Basalt. The continent is markedly asymmetrical, and the Eastern Ghats are younger and lower than the Western Ghats. The escarpment has a total length of over 1500 km and is seldom more that 60 km from the coast. In the south the escarpment is cut into

Precambrian rocks, but in the north across Cretaceous volcanics of the Deccan Traps.

The eastern highlands of Australia

Most maps of Australia show 'The Great Dividing Range' running inland of the eastern coast. There is a Great Divide separating drainage going to the Pacific from that going west, but for much of its length the divide crosses a nearly featureless plateau. Much more spectacular is the Great Escarpment. This huge landform runs for thousands of kilometres and often approaches a thousand metres high. When seen from the coast it looks like a 'range' but beyond it lies a plateau. In Australia the Great Escarpment has been described by Ollier (1982a), Pain (1985), and Ollier and Stevens (1989).

It is generally thought that the eastern highlands uplift occurred around the end of the Cretaceous and the start of the Tertiary, as sea-floor spreading of the Tasman Sea and the creation of the eastern Australian coast dates back to 80 million years ago. Erosion grading to the new coast and the new continental margin made valleys which coalesced to form the Great Escarpment, which continued to retreat. In numerous places it intersects volcanic flows around 30 million years old. Some younger volcanoes such as Tweed Volcano were erupted about 20 million years ago on the plains created by retreat of the escarpment, and are now considerably eroded themselves. Only one lava flow goes over the escarpment, and that is about 3 million years old. The escarpment is a diachronic feature, and is still eroding. It may have very nearly reached its present position some time ago, and it seems that such escarpments advanced very rapidly across soft rocks and have now reached hard rocks across which they are advancing very slowly.

The southern part of the Great Escarpment is related to the opening of the Tasman Sea. The northern, Queensland, section is related to an offshore rift which did not reach the stage of sea-floor spreading. In places the uplift of the plateau can be dated by sediment still covering the plateau, and some parts result from post-Cretaceous uplift. The escarpment can be dated by volcanoes it intersects (the Great Escarpment cuts across the 19 million-year-old Ebor volcano, so here is post-19 million years); by lava flows that run over the escarpment (a 3 million-year-old lava flow runs over the escarpment near Innisfail (Figure 8.11) so here the escarpment was in existence at that time, and has not retreated very far since); and volcanoes erupted after the passage of the escarpment (Tweed Volcano was erupted on the coastal plain about 25 million years ago after retreat of the escarpment). The escarpment is undoubtedly diachronic.

'What caused the uplift of the eastern highlands?' may not be an appropriate question according to Ollier (1992). Assembling evidence of widespread reversal of rivers along the east coast of Australia, he concluded that the rivers formerly flowed from Pacifica (land to the east of Australia) to the Great Artesian Basin, and when the Tasman Sea opened, the plateau was downwarped to the coast, reversing the rivers and creating a watershed (the Great Divide) on a plateau that was already high.

Greenland and Norway

As they are parts of the same Caledonian orogen broken apart by the sea-floor spreading of the Norwegian–Greenland Sea, the Lofoten–Vesterålen (Norway) and the Scoresby Sun area (Greenland) have comparable morphotectonics. Both these mountainous areas belong to the steep side of long marginal bulges of Tertiary age (Peulvast, 1988).

The Appalachians

It has been suggested that the Appalachians might be an example of a marginal swell and the Blue Ridge (an escarpment) might be a Great Escarpment. Battiau-Queney (1988) believes this is not so, because the Blue Ridge has been in about its present position since Early Mesozoic time, and the Piedmont erosional surface could not be formed by backwearing of the Blue Ridge scarp.

The Parakaima Mountains

In the northern part of South America, in Brazil, Venezuela and the Guyanas, lies a huge plateau known as the Parakaima Mountains. The plateau is underlain by horizontal Cretaceous marine strata and results from simple vertical uplift, obviously post-Cretaceous. Since then it has been attacked by erosion. On the northern side is a Great Escarpment, deeply embayed and sometimes isolating individual plateau remnants. The name Parakaima Mountains appears to be applied to both the plateaux and the bounding escarpment.

The Serro do Mar (Brazil)

The Brazilian Plateau is bounded on the Atlantic side by a Great Escarpment, which has been given various local names. The largest section is the Serra do Mar, which extends 800 km with a maximum height of 2245 m (Maak, 1969). It separates a coastal strip from the interior plateau, and all major streams drain westwards on the gentle inland slope to the Parana. There is little to date the uplift of the plateau, but the escarpment is presumably related to the opening of the South Atlantic, which started in the Jurassic, and is younger than the Parana Basalts, which are of Triassic age.

E. Alpine mountains

Alpine mountains in a geographical sense are those resembling in grandeur the European Alps. The Alpine orogeny refers to orogenic events in the Tertiary.

Alpine tectonics refers to the tectonics, regardless of age, characterized in their internal parts by plastic folding and plutonism, and in their external parts by lateral movements which have produced nappes, thrust sheets and closely crowded folds. In this chapter I shall use Alpine Mountains to mean actual mountains characterized by alpine tectonics.

The causes of alpine structures are controversial. The conventional explanation at present is compression at convergent plate boundaries, and Bates and Jackson (1987) even build this into their definition of alpinotype tectonics. Because the deformation is confined to an upper sheet without deformation of underlying basement, this kind of convergence is often called thin-skinned tectonics, as clearly the crust as a whole has not been compressed.

Others such as van Bemmelen (1954) and de Sitter (1964) explain the structures by gravity tectonics, a mechanism illustrated at length by De Jong and Scholten (1973). The basic objection to a lateral applied force, such as compression of plate edges, is that the strength of rock is not sufficiently great to transmit the force. All rocks would fail by flow or fracture if subjected to the forces of a moving plate. On the other hand a body force affects every atom within the body on which it is operating, not just at a point of contact as with an applied force. Gravity is the body force that affects rocks. Landslides are an obvious example, and many of the features of landslides are present in the much bigger nappes. For further discussion, see Ollier (1981, p.133).

The ranges of the Papuan Fold Belt are hard to explain by any mechanism other than gravity sliding. Deep drilling has shown that the folds are confined to the upper few kilometres (Figure 13.3), the basement is largely unaffected, and the main glide plane is within Jurassic strata. Where the strata could slide into basins they are deformed into complex folds and faults, and folding is greatest where a sliding mass came to rest against a basement obstacle. The sliding was consequent on vertical uplift of the central axis of country (Jenkins, 1974).

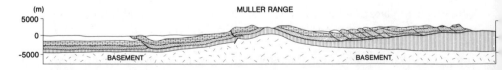

Figure 13.3 Cross-section of the Muller Range, part of the Papuan fold belt, Papua New Guinea. Note that the basement is not folded, nor the lower part of the Jurassic strata (blank), but the upper strata are deformed into complex folds and faults. (Simplified from Jenkins, 1974.)

Van Bemmelen emphasizes what he calls the bicausality of mountain-building — uplift is not caused by the same force that makes nappes. Basically he envisages broad uplift, making strata unstable so they slide down from geotumours into neighbouring depressions, where they pile up as nappes. But the location of uplift migrates and 'Finally the piles of nappes are overtaken by the forward shifting orogenic uplift and are exposed in lofty mountain ranges.'

The Apennines of Italy consist of a variety of arcs, some of them forming the mountain chain and some of them submerged beneath the Po Plain and the Adriatic Sea. Clearly the arc-forming process is not necessarily a mountain-forming process. Figure 13.4 shows the Apennine nappes approaching the Po Basin from the south, and the nappes of the Southern Alps approaching from the north, apparently on a collision course. But by no means does the apparent collision give rise to compression and mountain-building, and the Po Basin has been subsiding since the early Tertiary. According to Locardi (1988) the structure of the Apennine mountain belt is that of a succession of arcs corresponding to single thrust sheets. Each arc is bounded by transcurrent faults. They were formed in Oligocene to Middle Miocene times as a result of rotation of the Italian continental block meeting resistance from the rigid Adriatic foreland. On the other hand, the Calabria–Sicily arc is interpreted by Wezel (1982a) to be due to Plio-Quaternary vertical movement superimposed on structures formed by earlier compression.

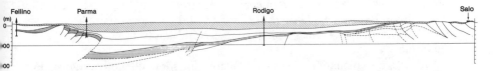

Figure 13.4 Cross-section of the Po Valley, showing Apennine thrusts moving in from the south, and thrusts from the Southern Alps moving in from the north, but instead of compression in the middle there is subsidence. The upper shaded layer is Quaternary sediment. The lower shaded layer is Upper Miocene sediment. The bedrock is Mesozoic.

The Tatra Mountains of Poland contain crystalline rocks which have been uplifted, according to fission track dating, about 15 million years ago. The general erosional dissection of the mountains was in Late Miocene and Pliocene times, and the glacial Pleistocene was only responsible for only minor landscape modification of the already dissected mountains (Glazek, 1989).

The European Alps have long been controversial so far as the origin of their structures is concerned, but their geomorphology is not so controversial, and not so old as in other mountain areas. Trumpy (1980) summarized the story of the Swiss Alps. The climax of the orogeny occurred at the end of the Eocene, about 40 million years ago, accompanied by volcanism. At the same time rifting of the foreland started in the Rhine Graben. The Alpine Foreland is closely linked to the uplift of the Alps. It was formed as the pre-Alpine depression simultaneously with the uplift of the Alps in the Upper Eocene and was filled with denudational and erosional debris, the Molasse, eroded from the rising mountains. The 3 km-thick Molasse consists of sediments that can be correlated with phases of uplift and erosion (Fischer, 1989).

The Alps were uplifted during the Oligocene, some granites emplaced, both strike-slip and dip-slip faults followed the main deformation. Thick sediments accumulated in the Molasse Basin. During the Oligocene and Miocene the Helvetic nappes were added to the Alpine pile, and the longitudinal furrows of the Alps developed. At the turn from Miocene to Pliocene the Jura Mountain folding occurred, which is so far the last act of deformation. In Pliocene times the Alps were worn down to a chain of low hills. The Swiss Alps of today were produced by strong uplift in the Pleistocene and strong glaciation.

F. Exotic terranes

Thorson (in Muhs et al., 1987) writes that

Much of Alaska is a mosaic of lithologically distinct tectono-stratigraphic terranes that have few equivalents in western North America. Consequently, considering this region as an extension of the Pacific Coast and Mountain System in the lower United States is a physiographic, and perhaps untenable, over-simplification.

This is merely emphasizing the point that if the exotic terrane hypothesis is valid, there may not be a common cause for physiographic features (including mountains) which are now adjacent. Similar reasoning could be applied to the mountains of Papua New Guinea or Irian Jaya if the exotic terrane hypothesis described on p.114 is true, and the same could apply to coastal ranges of California.

Alternative explanations may be available for some alleged terranes. For instance the Canadian Coast Plutonic Complex has been interpreted as being emplaced by displacement of the Canadian Cordillera by 2000 to 2500 km from the south, but Butler et al. (1989) suggested that regional tilt of 30 degress could explain palaeomagnetic anomalies better.

G. Domes

Broad domes may give rise to highland that is dissected into mountains. At the broadest level this is simply the cymatogenic uplift of King that gives rise to most mountains by plateau formation, but some domes are more local. The Lake District of England has already been described (p.35, Chapter 4). The doming took place in the Late Cretaceous or Early Tertiary, and landscapes essentially date back to that time.

A large dome from the Klamath Mountains of California and Oregon is described by Mortimer and Coleman (1985). The dome covers 70 000 km² and uplift at the centre is about 7 km. Regional doming is indicated by sub-annular outcrops of pre-Cretaceous rocks, radial high-angle faults, tilted Jurassic plutons, tilted Cretaceous to Miocene strata, and geomorphic features including Miocene fluvial deposits up to 850 m above the nearest modern major river. The Klamath River has an incised antecedent course where it crosses the dome, so the river pre-dates the dome. Doming occurred between 14 and 5 million years ago.

Chapter 14
Time, landforms and geomorphic theory

This book so far has described various features of landforms and landscapes from around the world. A picture has emerged of many ancient landforms, and a long time-scale that is required to explain the landscapes of today. Now I wish to see how this ties in with accepted ideas, theories and paradigms in geomorphology, and in earth science generally.

Some paradigms of geomorphology

Cyclical theories

Cyclical theories suggest that regions can be eroded until they become relatively flat, that tectonic uplift turns a plain into a plateau, and then renewed erosion once again wears the region down to a plain. The cycle can be repeated many times, interrupted, or modified by climatic change. There are variations on the theme, with Davisian peneplains and slope decline, King's pediplanes and parallel slope retreat, Crickmay's panplains formed by lateral erosion by rivers, and many more. All these cyclical theories have limitations in landscape explanation.

In the first place geomorphologists cannot agree on the existence, number or origin of erosion surfaces, and if they cannot agree on the field evidence there is no chance of agreement on explanations. The existence of multiple planation surfaces in many parts of the world indicates a succession of planation events, but we cannot be sure that erosion surfaces can be correlated or that they represent cycles rather than a succession of different events.

A long time-scale is brought into the argument by Ollier (1992) in a study of the Great Divide in Australia. The common name, Great Dividing Range, is a misnomer, but does suggest that there is a sharp range on the divide, and in Davisian terms this should become ever flatter as time goes on. In reality the divide started as a broad palaeoplain, in many areas dating back to the Triassic. Erosion affected the flanks, making a Great Escarpment on the steeper ocean-facing side, and a less continuous series of erosive effects on the less steep inland side. The palaeoplain was first divided into isolated high plains, and in a few places the opposing escarpments have met to produce a sharp ridge or 'Dividing Range' (Figure 14.1). This is roughly equivalent to the first half of a Davisian cycle, yet it has taken half of Phanerozoic time to be achieved in south-east Australia.

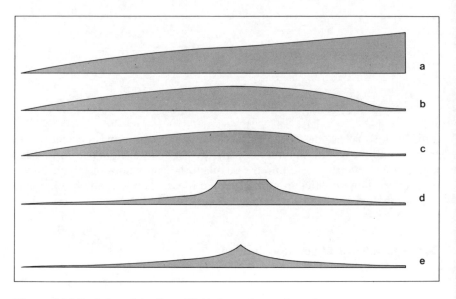

Figure 14.1 Evolution of the Great Divide in south-east Australia.

a. Initial palaeoplain sloping from the Tasman Divide in pre-Upper Cretaceous times.
b. Downwarp of the palaeoplain to the coast, forming an initial divide. Offshore the sunken palaeoplain is the base on which offshore sediments are deposited (the breakup unconformity).
c. Formation and retreat of the Great Escarpment facing the coast. Much of the Great Divide in New South Wales is in this stage of landscape evolution.
d. Retreat of slopes from the coast and inland reduces the palaeoplain to isolated High Plains, common on the eastern Victoria Divide.
e. Continued retreat of escarpments consumes the High Plains and produces a sharp ridge divide, as in much of eastern Victoria. The original palaeoplain was formed in the Triassic, so the landscape on a broad scale has progressed through half a Davisian cycle in about 250 million years.

Structural consideration can be brought into discussion of the cycle concept, as by Battiau-Queney (1989a and b). She analyzed the structure of the British Isles and proposed that the crust is very heterogeneous, due to the accretion of many terranes. Each crustal block has its own isostatic equilibrium, and adjacent hinges are separated by deep-rooted faults or 'crucial hinges'. Some of the blocks can be easily reactivated, and at a regional scale, long-term landscape evolution can lead to an increased relief instead of the planation surface predicted by conventional cyclic theories. In her own words:

The cycle of erosion is a model of evolution which implies a series of sequential landform changes through time. Successive stages are organized in a progression leading to a planation surface. The general base-level exercises the main control on this evolution. This model is incompatible with an heterogeneous crust: the state of stress applied on it is closely controlled by all the discontinuities inherited from the former geologic history. The long-term landform evolution depends mainly on the crustal properties. Old palaeoforms can survive a very long time, close to areas which have suffered rapid changes in the last million years.

Thus she shares Crickmay's view, described later, about the great differences in rates of geomorphic activity. Furthermore, she concludes that 'the lowest planation surfaces are not necessarily the youngest.' In the British Isles, she claims, 'the oldest palaeosurfaces are found in low areas which were not notably uplifting or subsiding since the late Palaeozoic. Moreover, the morphologic evolution does not tend to a planation surface.' What is abundantly clear is that the basic tectonic assumption of the Davisian cycle, that tectonic movement is upwards, periodic and sudden cannot be accepted.

The systems approach

Chorley (1962) maintained that if geomorphologists thought in terms of an 'open system', they would attain more fruitful ideas and profitable concepts than if they thought in the 'closed system' of W.M. Davis. The 'systems approach' has had a long run and found a place in many textbooks, so there is a need to consider if it has any value in long-term landscape studies, or indeed any value at all.

A closed system has no transfer of energy or material across the boundaries of the system: an open system has transfer of energy and/or material across the boundaries of the system. If we consider a landscape experiencing erosion according to the Davisian model, there is input of energy from precipitation, and there is also removal or eroded sediment. Thus there is transfer of both energy and material across the boundaries of the system, which is therefore an open system, and was never thought to be otherwise. In fact all evolving landscapes are open systems, and it is simply an injustice to Davis to misrepresent his erosion cycle as a closed system.

It is interesting to see how the misuse of the 'closed system' grew in the literature. In 1950 Strahler wrote, 'A graded drainage system is perhaps best described as an open system in a steady state (Von Bertalanffy, 1950) which differs from a closed system in equilibrium in that the open system has import and export of components.' This is uncontroversial, even if unhelpful.

In the early part of his 1962 paper Chorley cautiously says that some aspects of Davisian geomorphology can be considered analogous to some aspects of closed systems, for instance: 'Thus, one can see the Davis' view of landscape development contains certain elements of closed system thinking' (Chorley, 1962, p.2), and: 'This again enables one to draw striking analogies between closed-system thinking and the historical approach to landform study which was proposed by Davis' (Chorley, 1962, p.3). Later the analogy gives way to direct indentification: 'Davisian (closed system) thinking is instinctively opposed'... (Chorley, 1962, p.8).

When Chorley's article is summarized by Hack we find:

Chorley (1962) suggested that the concept (Davisian) in reality treats an eroding landscape as a closed system in which the energy in the system becomes less as erosion proceeds towards base level and finally in the last stages approaches zero.

Or as Higgins (1975) expressed it:

In these terms Davis' 'Geographical Cycle' is identified as a *closed system*, which receives all of its energy at the outset and only reaches equilibrium and a steady state at the end, when the machine has run down, so to speak.

It might be argued (not successfully, I think) that a landscape gets its potential energy initially from tectonic uplift. In reality its erosive energy comes from wind and rain and runoff which extend throughout the life of the landscape. The false terminology is also sometimes used in discussing dynamic equilibrium, such as 'The basic premise for the system (Hack, 1960, p.81) is that the landscape and the processes that form it are part of an open system which is in a steady state of balance' (Palmquist, 1975).

In reality all landscapes are open systems, and all models of landscape evolution conceive them as such. In this case it is hard to see how the benefits of 'open system thinking' should suddenly accrue when one abandons the Davisian cycle. It seems to me that the open systems approach has contributed little to the progress of geomorphology, and I have dealt with it at some length only because it was used as a basis for the next paradigm, dynamic equilibrium.

Dynamic equilibrium

Although it is commonly traced back to Gilbert (1877), this model in its modern form was proposed by Hack (1960). The situation envisaged by Hack and others and called by them 'dynamic equilibrium' is one in which weathering, removal by erosion and addition by deposition are so balanced on a slope (or other landform) that there is no change in form through time. Thus a point on a hillslope may receive debris from upslope, produce debris in place, and lose debris to downslope sites, and if the amounts are 'balanced' the spot remains topographically unaltered. A whole hillslope profile may be in this sort of balance (except the top, where addition from above is impossible). Furthermore, if uplift rate of the land is exactly equal to the downwasting rate, then an assemblage of slopes in a landscape remain constant in form through a long period of time. That is a very big 'if', and most unlikely to prevail for very long anywhere.

In dynamic equilibrium theory, it is assumed that within a single erosional system all elements of topography are mutually adjusted so that they are downwasting at the same rate. The forms and processes are in a steady state of balance, and may be considered as time-independent. Equilibrium is achieved when all the slopes in a drainage basin are mutually adjusted to a common erosion rate. So long as uplift is maintained and the rivers do not reach base level, the landscape can maintain the same form indefinitely, even though erosion is continuous. Such a landscape could be considered 'timeless', for the topography would be indentical at widely separate times.

In terms of cyclical theories, the dynamical equilibrium theory only applies to the middle stage (maturity) when the landscape consists of all slopes, with no remnants of ancient peneplains and no development of new forms at a new base level. Hack denied that remnants of old erosion surfaces are present, though Bretz (1962) demonstrated one in Hack's own type area. However, even Hack himself admitted that when a new base level is reached, landforms must change. He wrote:

If the [rates of uplift and erosion] change, however, then the state of balance or equilibrium constant must change. The topography then undergoes an evolution from one form to another. Such an evolution might occur if diastrophic forces ceased to exert their influence, in which case the relief would gradually lower.

In other words the landscape forms evolve except in special conditions! Even the originator of the theory admits that dynamic equilibrium applies only to special-case landscapes.

But then Hack tries to keep evolving landscape within the realm of dynamic equilibrium by claiming:

Nevertheless as long as diastrophic forces operate gradually enough so that a balance can be maintained by erosive processes, then the topography will remain in a state of balance even though it may be evolving from one form to another.

He seemed to admit grudgingly that relict landforms exist when he wrote, 'If, however, sudden diastrophic movements occur, relict landforms may be preserved in the topography until a new steady state is achieved.'

Dynamic equilibrium may conceivably apply in some situations over some time span, but I believe the many examples in this book show change over time to be prevalent, and that dynamic equilibrium is quite rare, confined essentially to all-slope landscapes (mature landscapes of Davis). I find the concept unhelpful in long-term landscape studies, but some workers seem to find it worth discussing, including the following.

Suppe (1987) after describing the geology of Taiwan, concluded that topography in a tectonically active area reaches steady state in just over a million years. Slates and phillites form and reach the surface by erosion within 2 to 3 million years. Sediments derived from the mountains are trapped adjacent to the mountain belt at rates of 1 000 to 10 000 B. The mountainous relief has decayed to a tenth of its initial value in less than a million years. Unfortunately, Suppe's account, while rich in stratigraphic information and plate tectonic explanation, does not contain enough detail of the landforms to test the validity of these conclusions.

Cerveny et al. (1988) from their studies related to the Himalayas (see p.192, Chapter 13), concluded that the Himalayan landscape on a broad scale has been in a relatively steady state for at least 18 million years. The significant feature of their conclusions lies in the words 'broad scale' and they do not envisage individual hills and valleys remaining in a steady state.

Schumm and Lichty (1965) proposed that the difference between steady-state, dynamic equilibrium attitudes and cyclic attitudes is a matter of scale. The size of the landscape involved and the amount of time considered affect the observer's viewpoint. The more specific we become, the shorter is the time span with which we deal and the smaller is the space we can consider. They suggest that landform evolution can be considered during three time spans of different duration: cyclic, graded and steady. The cyclic time span encompasses a major period of geologic time, perhaps involving an erosion cycle. Over this time span the system will change because of significant removal of material by erosion and possibly by tectonic movements. Graded time refers to a short period of cyclic time when any slight progressive change in landforms is masked by fluctuations about the average values. Steady time refers to a span in which static equilibrium may exist and landforms are truly independent because they do not change. For example, a sand bar in a river may gain sand grains from upstream and lose them downstream, but retain the same shape.

Clearly anything to do with long-term landscape evolution is in 'cyclic time', but this is an unfortunate term because it suggests automatically that the cyclic approach is correct. Schumm and Lichty are correct in showing that concepts of steadiness, grade, equilibrium or 'dynamic equilibrium' can only refer to landscape evolution over a short time, but some other term is needed for 'cyclic time'. Perhaps long-term landscape evolution is all that is required.

Starkel (1987), after analysing the Carpathian Mountains in Poland, concluded that the relative importance of inherited forms depends to a large extent on the spatial scale being considered, with the largest forms being related primarily to Miocene tectonics, and the smallest features, like active river channels, being in equilibrium with current hydrological conditions. This situation applies to the Polish Carpathians specifically, but the many examples in this book show a far greater range in historical variation.

Climatic geomorphology

Some geomorphologists, notably Budel (1982), thought that climate was the dominant factor in landscape evolution, but that modern landscapes are complicated by inheritance of features created in former climates and preserved in part to the present day. This is certainly true to some extent, and is most obvious in areas that were glaciated in the last ice age and are now experiencing fluvial erosion

When so much of the landscape of today is of an age measured in geological periods, the landscapes have inevitably passed through a variety of climates in the past. Yet the mark these climates leave is not always overwhelming and is often negligible. How else can we account for many different interpretations of the same place? Deep weathering in Europe was once attributed routinely to Interglacial weathering, then to Tertiary weathering, and it is now known that some at least is Mesozoic. Some landforms are good climatic indicators, such as glacial landforms, desert dunes and coral reefs. Others, such as inselbergs and tafoni have been used as alleged indicators but seem to have failed the test, and fluvial landforms and slopes have not proved to be useful indicators of past climates. The limits of climatic geomorphology are further discussed by Stoddart (1969).

Battiau-Queney (1987) pointed out that some features of Tertiary landscapes are preserved in the British Isles, even in glaciated areas. These include remnants of deep weathering profiles, palaeokarst, pediments, tors and inselbergs. She emphasized that an important factor in the recognition of these old features is that there has been a major climatic change, so the tropical features of the Early Tertiary are distinct from later, non-tropical landforms.

Etchplanation

According to Summerfield and Thomas (1987), the overwhelming opinion about the nature of landscape modification at the present time is for landscape lowering associated with the formation and removal of deep weathering profiles. This puts an emphasis on a kind of etchplanation, but also hints at steady-state evolution of parts of the landscape and preservation of others — we could only detect it if the lowering was localized and parts of earlier landscapes were preserved. I do not know of any basis for the 'overwhelming opinion', but my own view is that etchplanation is understood and thought widespread by only a minority of geomorphologists at present.

The hypothesis of unequal activity

C.H. Crickmay will be remembered for two contributions to geomorphology. The first is the idea of panplanation, in which lateral erosion causes planation surfaces in contrast to slope decline in the Davisian peneplain scene, and parallel slope retreat in the pediplanation theory of King. The second, and more important, was what he called the 'hypothesis of unequal activity' (Crickmay, 1975). He presented a range of evidence to show that geomorphic activity may vary by a factor of a million. The preservation of the many ancient landforms described in this book suggests this is a true and necessary concept.

Historical geomorphology

This is the old, traditional view of landforms and is concerned with how they were formed, what stages they went through, and when it all happened. The cyclic paradigm is an example of historical geomorphology, though a cyclic approach is not necessary for the study of historical geomorphology. For a long time historical studies were called denudation chronology. This term has fallen into disuse because geomorphic history includes not only denudation, but also depositional and tectonic aspects of landscape evolution. This is seen by some as the geological approach to geomorphology, and Thornbury (1969) had for his Fundamental Concept 10: 'Geomorphology, although concerned primarily with present-day landscapes, attains its maximum usefulness by historical extension.'

Chorley (1962) wote: 'classical' geomorphologists have retreated into restricted historical studies of regional form elements, whereas, similarly, quantitative workers have often withdrawn into restricted empirical and theoretical studies based on processes.' It is my belief that there is no question of 'withdrawal' or 'retreat', but that studies of geomorphic history are a valuable contribution to science, as indeed are process studies.

Some extraordinary arguments have been used to discredit the historical approach to geomorphology. Chorley (1965, p.34) pointed out, as if it were shameful, that denudation chronology is really a branch of historical geology. This is true, of course, and there is nothing wrong with it. The same author (1965, p.148), still worried about the geological affinity of denudation chronology, claimed that it is weak because it is not purely morphological, although later (p.151) he wrote, 'most purely morphological evidence is so ambiguous that theory feeds readily on preconception.' He contrasts (p.149) two different approaches to landforms as the 'historical hangover' and 'dynamic equilibrium'. This emotive rhetoric does not avoid observed facts that many landforms are due to former events and processes and so have 'historical hangover', though some may indeed be in equilibrium with present conditions. Only by the study of real geomorphic history can we decide what is really going on. It is futile to engage in 'research by debate' (one of Chorley's (1965, p.31) criticisms of the 'old geomorphology'!).

Chorley (1978) discussed several phases in the history of the study of geomorphology, including teleological, taxonomic, functional and historical. Of the historical approach he wrote: 'Historical. The explanation of landforms in terms of a narrative reconstruction of a series of events assumed to have led up to the present. Such palaeogeographical speculations are often found to rely on allegorical or metaphysical underpinning.' How this opinion is reached is not clear, but presumably all historical geology and stratigraphy must also be based on 'allegorical or metaphysical underpinning'. Such verbal and metaphysical acrobatics should not obscure the fact that the past is there to be studied, and it is not possible to understand the landforms of today without knowing something of their past.

Economic geology and ancient landscapes

Plate tectonics is a well-known revolution in earth science. A lesser-known revolution is occurring in economic geology. The dominant idea in ore genesis for many decades was that metals were brought from depth in 'emanations' or 'ore-forming fluids', probably from granites intruded beneath. This idea has been supported by observation of ore-forming 'black smokers' on the sea-floor, but modern ideas have much more emphasis on near surface enrichment in sediments and in the regolith. King (1989) has presented an account of this revolution, with many incidental details. He contrasts 'primary' and ascensionist ores with those that are in some way surficial and descensionist. The latter must relate to past or present landscapes.

In Kalgoorlie, for example, there was a great deal of secondary enrichment of the shallower ore (gold), and on the margins of the Golden Mile many small shoots have been stoped to 100 feet or so and then abandoned due to impoverishment. King concluded that secondary enrichment is general in the Kalgoorlie area. This leaves open the question of the primary origin of the gold.

More significant evidence comes from finding major ore deposits, such as in McArthur River in northern Australia, a lead zinc deposit in Precambrian sediments, stratiform, unfolded, unmetamorphosed and devoid of igneous associations. Such a deposit could then be envisaged as a primitive, simple, bedded precursor of the now much-altered Broken Hill ore deposit.

In the Mount Isa-Cloncurry region an economic survey led to the conclusion that the region has been exposed to surficial effects 'going beyond ordinary weathering', possibly in the Precambrian, probably in pre-Cretaceous time as well as more recently. This fits in with the geomorphic indicators of great landscape age, such as the Cambrian river terraces of the Davenport Ranges (p.22).

King gives numerous examples of deep weathering, much of which must be attributed to time well before the present. Weathering effects include:

1. oxidation, leaching and secondary enrichment in sulphide ores to depths of 150 to 200 m at Cobar and Broken Hill (New South Wales), Gunpowder (Queensland) and Tennant Creek (Northern Territory);
2. oxidation to 350 m at Wingellina (South Australia);
3. oxidation and leaching of copper-bearing carbonates to 800 m at Mount Isa (Queensland), that is to 250 m below sea-level;
4. leaching and enrichment to 250 m at Mount Goldsworthy (Western Australia), that is to 170 m below sea-level.

Even when ore deposits are connected with ancient landscapes not immediately related to those of the present day, the geomorphic associations of ore remain interesting. Examples include:

1. The lead–zinc deposits of the Carpentaria region (northern Australia) were formed in a lagoonal, emergent sedimentary environment.
2. The iron–copper–uranium deposit of Olympic Dam (South Australia) is part of a ferruginous breccia in a graben beneath a major unconformity.
3. The ore of Mount Lyell (Tasmania) lies at the base of a ferruginous breccia in a rift beneath an unconformity.

Other examples of oxidized and sulphide zones are given by Ollier (1984). The oxidized zone reaches 600 m and deeper at Kennecot, Alaska; Tintic, Utah; Zambia; and at the Lonely Mine, Zimbabwe it reaches a depth of

900 m. The sulphide zone is 150 m at Ely, Nevada; 300 m at Butte, Montana and at Morenci, Arizona; and 420 m at Bingham, Utah.

The boundary between the oxidized and sulphide zones corresponds to the position of the water table at the time of ore genesis, but that level may now be different. At Cripple Creek, Colorado, oxidized ores are found 60 m below water-level; in Zambia oxidized ore is submerged up to 600 m. In other places including Bingham, Utah; Bisbee, Arizona; and Rio Tinto, Spain, the sulphide zone is now stranded above the water table.

Over the past twenty-five years many workers have tried to relate mineral deposits to plate tectonics, with some apparent success. The many limitations to the association have been marshalled by Sangster (1979). Virtually all efforts to relate plate tectonics to mineral deposits adopt the same approach. Major plate boundaries such as spreading sites, subduction zones, collision zones and transform faults are first described in general terms, and then mineral deposits thought to be associated with each plate element are described. This automatically produces a list of plate-related deposits which seems impressive in the number and diversity of the economic minerals. But when the list is considered against the total list of economic deposits, it appears that apart from small and inconsequential deposits, only porphyry coppers and volcanogenic sulphides have good plate tectonic links.

In contrast, it can be shown that many deposits cannot be related to plate tectonic sites and processes, including red-bed copper and uranium, Proterozoic banded iron formations, *Kupferschiefer* copper, Mississippi Valley type lead and zinc deposits, stratiform barite and phosphate, and the many ores with a major geomorphic component formed by weathering, erosion and deposition. The fact that many mineral deposits are formed on the continental plate and not at the margins suggests that plate tectonics is only of limited use in understanding the origin and distribution of mineral deposits. A compilation of ore deposits in relation to geomorphic setting rather than plate tectonic setting would be a very useful exercise that has not yet been seriously attempted.

The age control on ore formations is sometimes not very firmly established, but they undoubtedly bespeak a long time-scale. There is need for greater co-operation between economic geologists and geomorphologists who can handle geological time. The combination has great potential for important discoveries.

Tectonic geomorphology (morphotectonics)

Plate tectonics

Plate tectonics, described in Chapter 8, is the ruling theory in earth science today. It provides a unifying system that accounts for many of the major features of the earth, both in geology and in geophysics. The very weakness of plate tectonics, is that the power of explanation is too great. Subduction, for example, may be called upon to create an island arc, a deep sea trench, or a mountain range; it may lead to sediments being scraped off a down-going slab, or subduction of the sediments under the continent. Tectonic collision may be invoked to make folds by compression, or back-arc basins by tension. 'The fact that the subduction hypothesis can account for both uplift and subsidence, compression and tension, means that it has too many degrees of freedom. It can account for opposite effects and is not testable' (Ollier and Pain, 1988). The eminent Swiss stratigrapher Trumpy wrote of himself in 1980: 'He is fascinated by plate tectonics but extremely diffident of ready-made explanations.' This seems to be a healthy attitude.

Most of the input into plate tectonic speculation appears to come from geophysicists, who have models which may be elegant mathematically but do not take landforms into account at all. Plate tectonic theories of mountains do not normally consider geomorphic evidence. For instance, in the great amount of literature on the plate tectonic evolution of the Himalayas by collision of India and Asia, possibly pushing India under Asia, and so forth, the evidence of the antecedent rivers (p.32) is never mentioned, though it has been in the literature since 1937. Geomorphology should have real input into plate tectonics, but until more geomorphologists are working on long-term landscape evolution, the present situation is likely to prevail.

Exotic terranes

The possibility of exotic terranes in a landscape seems to have been hardly considered from the point of geomorphic theory, but it has great ramifications. Any regional geomorphic studies must in future be on guard for the possibility of exotic terranes, because the geomorphology of different terranes will be different. It is necessary to decipher the geomorphology of each terrane, and geomorphic consequences of docking, such as formation of rivers along sutures. The long-term landscape evolution will include the history of individual terranes, and the evolution of the amalgamated unit after docking, when at least some of the geomorphic features will be common to all. The example of the British Isles described by Battiau-Queney (1989) described on p.200 shows how very different a geomorphic history can be if an exotic terrane approach is used.

Active margins

In plate tectonics, collision boundaries are where most of the action is, and where geomorphology should be making great advances. Unfortunately, as explained earlier there is too much freedom of interpretation at collisional sites, and many of the 'explanations' offered for the morphotectonics at active margins are really only rationalizations. In a volume devoted to the geomorphology of active margins (Williams, 1988), it was disappointing to see that in most papers the plate tectonic basis was assumed and not demonstrated, and frequently the geomorphology could have been presented just as well, or better, without the complication of assumed collision.

It is worth remembering that some collision margins do not behave as they should. Where the north-moving nappes of the Apennines meet the south-moving nappes of the Alps there is no compression, but instead the Po Valley depression, which has been sinking for 40 million years. In Papua New Guinea, a classic collision site where the Pacific Plate collides with the Australian Plate, there is abundant evidence for uplift, tension and horizontal movement, but not for compression. Even in the Himalayas there are tensional features as well as the well-known nappes.

Passive margins

Summerfield (1987) claimed the geomorphological evolution of passive margins must be seen in the context of four primary factors:

1.   the timing and spatial extent of the rift-related uplift along the margin;
2.   the nature of the thermally-driven post-rift subsidence;
3.   the isostatic response of the margin to denudational unloading (largely through escarpment retreat);
4.   the flexure of the margin associated with sediment loading offshore.

As explained in Chapter 8 none of these factors can be assumed, but must be tested against real-world data. In eastern Australia, for instance, the uplift may not be thermally driven, and there is no sign of isostatic response despite large-scale escarpment retreat. The study of passive margins has made great advances in the past ten years, and the relationship between morphotectonics and associated sedimentary basins enables a good time-scale to be derived. It is quite clear that the time-scale has to be long.

Uniformitarianism

In simple language, uniformitarianism means that 'the present is the key to the past.' Bates and Jackson (1987) define it more formally thus:

The fundamental principle or doctrine that geologic processes and natural laws now operating to modify the Earth's crust have acted in the same regular manner and with essentially the same intensity throughout geologic time, and that past geologic events can be explained by phenomena and forces observable today.... The doctrine does not imply that all change is at a uniform rate, and does not exclude minor local catastrophes.

Uniformitarianism is widely accepted as *the* basic principle of geology, but the literature is riddled with misconceptions, misleading statements and false conclusions. To illustrate the extreme breadth and depth of the misconceptions, Shea (1982) framed the most pernicious and pervasive of them into a set called 'Twelve fallacies of uniformitarianism' (Table 14.1). There is no space to discuss the fallacies further here, but Shea claims that the only sound analysis of modern uniformitarianism is by Goodman (1967), and concludes that 'modern uniformitarianism has no substantive content — that is, it asserts nothing whatever about nature.'

Table 14.1 Twelve fallacies of uniformitarianism

1. Uniformitarianism is unique to geology.
2. Uniformitarianism was first conceived by James Hutton.
3. Uniformitarianism was named by Charles Lyell, who established its definitive modern meaning.
4. Uniformitarianism should be called 'actualism' because it refers to the 'actual' or 'real' events and processes of Earth history.
5. Uniformitarianism holds that only currently acting processes operated during geologic time.
6. Uniformitarianism holds that the rates or intensities of processes are constant through time.
7. Uniformitarianism holds that only 'gradual', 'noncatastrophic' processes have occurred during Earth's history.
8. Uniformitarianism holds that conditions on earth have changed little through geological time.
9. Uniformitarianism holds that earth is very old.
10. Uniformitarianism is a theory or hypothesis and can be tested.
11. Uniformitarianism applies only as far back in history as present conditions existed and only to earth's surface or crust.
12. Uniformitarianism holds that the laws governing nature are constant through space and time.

Thomas and Summerfield (1987) wrote:

The twin processes of weathering and erosion have been the traditional concern of historical geomorphology for nearly a century, during which the approach has swung from the anthropogenic 'cycle of erosion' (Davis, 1899) to the open system in 'dynamic equilibrium' (Hack, 1960). Both concepts implicitly support the view that 'the present is the key to the past' in the earth sciences. But the understanding that earth environments and continental configurations have evolved through geological time, creating unique landscapes containing inherited and superimposed features ...limits the usefulness of this dictum.

From geological and geomorphic evidence we are quite sure that many things were different in the past. For example, before 250 million years ago there was no grassy cover to the landscape, so erosion rates were probably faster. Other things that were different in the past include latitude, altitude, distance to the sea, climate, ocean currents, wind systems, vegetation cover and soil-forming processes. More fundamentally, there may have been changes in day length, gravity, the earth's radius and the tilt of the earth's axis. We have to learn what the past was like from consistent internal evidence, not from comparison with places that are thought to be somewhat similar today. The study of long-term landscape evolution will help to decipher the past.

The effects of too much uniformitarianism in thinking are well expressed by Higgins (1975) who wrote:

To my mind — and this reflects a very biased judgment — the main reason that we still lack satisfactory theories of landform development is that in most geomorphic field studies investigators have misinterpreted the relationships between form and process because they have failed to recognize that in many parts of the world the gross forms of the landscape are relics formed by processes no longer operating there.

Evolutionary geomorphology

The earth is evolving, and landscapes are evolving too. Instead of seeking dynamical equilibrium, cycles, or even uniformitarianism, it might be more useful to see how landscapes actually evolve. The study of long-term landscape evolution may enable us to do this.

In the very different past the whole geomorphic system was also very different. In the Precambrian there may have been different morphotectonic systems, with ductile flow rather than brittle fracture and failure characteristic of later times (Wynne-Edwards, 1977). The early earth had a reducing atmosphere, and geomorphology would have changed considerably after an oxygen-rich atmosphere evolved.

The present era of continental drift started in about Jurassic times, and continues to the present. In earlier times the continents were united as Pangaea. Geomorphology on that super-continent would have been very different from that of today, resulting from the existence of vast inland areas at great distance from the sea, longer rivers, more inland deposition and very different climatic patterns. The breakup of Pangaea would have had important effects on each fragment, with rivers having shorter courses to the sea, rejuvenation and increased erosion on new continental edges, tectonic warping of new continental margins. The amalgam of continental fragments such as India or exotic terranes as in Papua New Guinea would bring together areas with very different geomorphic histories.

Each fragment of Pangaea would have its own distinct history, with many unique events such as formation of new plate edges, biological isolation and local evolution, and changes in latitude and climate. The geomorphology of each fragment must be seen on the long time-scale that is appropriate for continental drift, mountain-building and biological evolution. In this context some of the theories and fashions of geomorphology — process studies, dynamic equilibrium, and cyclical theories — appear to have limited importance or application. What we see is an evolutionary geomorphology, which is part of the story of an evolving earth.

However, many examples presented in this book relate to times before the Jurassic, and before the present phase of plate tectonics and continental drift. This geomorphology of ancient times provides basic information to be used in building up the story of the earth. Geomorphologists should not simply accept theories of global geology from geologists and geophysicists and apply them to landscapes — they have basic data that should be incorporated in the building of models of the earth on the long time-scale.

Epilogue

In this book I have presented a whole range of snippets of information about ancient landforms, often without any critical comment. Sometimes I have quoted different workers whose conclusions appear to clash, and have not tried to analyse or adjudicate. My aim has been to present a mass of evidence to show that geomorphology must be seen on a long time-scale, that evidence is available, and that useful conclusions can be drawn. I hope that geomorphologists will take up the challenge and devote themselves to the rewarding study of long-term landscape evolution, so that this book, which is mainly arguing a case, can be replaced by a systematic account of the subject. I also hope that geologists will realize the value of incorporating geomorphic data into their work, both in practical geology such as work on ore genesis, clays, or hazards, and the theoretical aspects such as the study of continental margins, sedimentary basins, and global tectonics.

References

Abbate, E., Bruni, P., Fazzuoli, M. and Sagri, M. 1986. The Gulf of Aden, continental margin of northern Somalia: Tertiary sedimentation, rifting and drifting. *Mem. Soc. Geol. It.*, 31, 427–45.

Ahnert, F. 1989. The major landform regions. *Catena*, Supp.15. 1–9.

Albritton, C.C., Brooks, J.E., Issawi, B. and Swedan, A. 1990. Origin of the Qattara Depression, Egypt. *Bull. Geol. Soc. Am.*, 102, 952–60.

Ambrose, J.W. 1964. Exhumed paleoplains of the Precambrian Shield of North America. *Am. J. Sci.*, 262, 817–57.

Andres, W. 1989. The central German Uplands. *Catena*, Supp.15, 25–44.

Andrews, E.C. 1910. Geographical unity of eastern Australia in Late and Post Tertiary time. *J. Proc. R. Soc. N.S.W.* 44, 420–80.

Artyushkov, E.V., Letnikov, F.A. and Ruzhich, V.V. 1990. The mechanism of formation of the Baikal Basin. *J. Geodynamics*, 11, 277–91.

Asmeron, Y., Snow, J.K. et al. 1990. Rapid uplift and crustal growth in extensional environments: an isotopic study from the Death Valley region, California. *Geology*, 18, 223–6.

Augustinus, P.C. In press. Reconstruction of former ice-caps and outlet flow patterns, Fiordland, New Zealand. *Geografiska Annaler.*

Auzende, J-M., Lafoy, Y. and Marsset, B. 1988. Recent geodynamic evolution of the north Fiji basin (southwest Pacific). *Geology*, 16, 925–9.

Baker, B.H. and Mitchell, J.G. 1978. Volcanic stratigraphy and geochronology of the Kedong-Olorgesailie area and the evolution of the South Kenya rift valley. *Jl. geol. Soc. Lond.*, 132, 476–84.

Baker, B.H., Mohr, P.A. and Williams, L.A.J. 1972. Geology of the eastern rift system of Africa. *Geol. Soc. Am.* Spec. Paper 136.

Bardossy, G. 1989. Bauxites. pp. 399–418 in Bosak et al., 1989, q.v.

Bardossy, G. and Aleva, G.J.J. 1990. Lateritic Bauxites. Developments in *Economic Geology* 27. Elsevier, Amsterdam.

Barnovsky, A.D. and Labar, W.J. 1989. Mid-Miocene (Barstovian) environmental and tectonic setting near Yellowstone Park, Wyoming and Montana. *Bull. Geol. Soc. Amer.*, 101, 1448–56.

Bates, R.G., Beck, M.E. and Burmester, R.F. 1981. Tectonic rotations in the Cascade Range of southern Washington. *Geology*, 9, 184–9.

Bates, R.L. and Jackson, J.A. 1987. *Glossary of Geology*. 3rd edn. American Geological Institute, Alexandria, Virginia.

Battiau-Queney, Y. 1984. The pre-glacial evolution of Wales. *Earth Surf. Proc. and Landforms*, 9, 229–52.

Battiau-Queney, Y. 1986. Buried palaeokarstic features in South Wales: examples from Vaynor and Cwar Yr Ystrad quarries, near Merthyr Tydfil, in Paterson, K. and Sweeting, M.M. (eds) 1986. *New Directions in Karst.* Geo Books, Norwich. 551–67.

Battiau-Queney, Y. 1987. Tertiary inheritance in the present landscape of the British Isles, in Gardiner, V. (ed.) *International Geomorphology 1986*, Part II. Wiley, London. 979–89.

Battiau-Queney, Y. 1988. Long term landform development of the Appalachian Piedmont (USA). *Geog. Annal.*, 70A, 369–74.

Battiau-Queney, Y. 1989a. The relationship between long-term landform development and the crust underlying the British Isles. *Rev. Geomoph. Dynamique*, 38, 1–15.

Battiau-Queney, Y. 1989b. Constraints from deep crustal structure on long-term landform development of the British Isles and eastern United States. *Geomorphology*, 2, 53–70.

Beckmann, G.G. 1983. Development of old landscapes and soils, in *Soils: an Australian viewpoint*, Division of Soils, CSIRO. 51–72. CSIRO, Melbourne; Academic Press, London.

Bemmelen, R.W. van. 1954. *Mountain Building.* Martinus Nijhoff, The Hague.

Benbow, M.C. 1990. Tertiary coastal dunes of the Eucla Basin, Australia. *Geomorphology*, 3, 9–29.

Bird, M.I. and Chivas, A.R. 1989. Stable-isotope geochronology of the Australian regolith. *Geochim. Cosmochim. Acta*, 53, 3229–56.

Bird, M.I. Chivas, A.R. and Andrew, A.S. 1989. A stable-isotope study of lateritic bauxites. *Geochim. Cosmochim. Acta*, 53, 1411–20.

Bird, M.I. Chivas, A.R. and McDougall, I. 1990. An isotopic study of surficial alunite in Australia. 2. Potassium-argon geochronology. *Chemical Geol.* Isotope Geoscience Section) 80, 133–45.

Bishop, P. 1985. Southeast Australian late Mesozoic and Cenozoic denudation rates: a test for late Tertiary increases in continental denudation. *Geology*, 13, 479–82.

Bishop, P. 1986. Horizontal stability of the Australian continental drainage divide in south central New South Wales during the Cainozoic. *Australian Journal of Earth Science*, 33, 295–307.

Bishop, P. 1988. The eastern highlands of Australia. *Progress in Physical Geography*, 12, 159–82.

Bishop, P., Hunt, P. and Schmidt, P.W. 1982. Limits to the age of the Lapstone Monocline, N.S.W. — a palaeomagnetic study. *J. Geol. Soc. Aust.* 29, 319–26.

Blanck, H.R. 1978. Fossil laterite on bedrock in Brooklyn, New York. *Geology*, 6, 21–4.

Bosak, P., Ford, D.C., Glazek, J. and Horacek, I. 1989. *Paleokarst.* Elsevier, Amsterdam.

Bourrouilh-Le Jan, F.G. 1989. The oceanic karsts: modern bauxite and phosphate ore deposits on the high carbonate islands of the Pacific Ocean, in Bosak et al. 1989, q.v. 443–71.

Bovis, M.J. 1987. The interior mountains and plateaus. in Graf, W.L. (ed.) q.v. 469–515.

Branagan, D.F. 1983. The Sydney Basin and its vanished sequence. *J. Geol. Soc. Aust.* 30, 75–84.

Branagan, D.F. and Pedram, H. 1990. The Lapstone structural complex, New South Wales. *Aust. J. Earth Sciences*, 37, 23–36.

Bremer, H. 1989. On the geomorphology of the South German scarplands. *Catena, Supp.* 15, 45–67.

Bretz, J.H. 1962. Dynamic equilibrium and the Ozark landforms. *Am. J. Sci.* 260, 427–38.

Briceno, H.O. and Schubert, C. 1990. Geomorphology of the Gran Sabana, Guayana Shield, southeastern Venezuela. *Geomorphology,* 3, 125–41.

Bristow, C.M. 1968. The derivation of the Tertiary sediments in the Petrockstow Basin, North Devon. *Proc. Ussher Soc.,* 2, 29–35.

Bristow, C.M. 1990. Ball clays, weathering and climate, in Zupan, A.W. and Maybin, A.H. (eds), *Proc. 24th Forum on the Geology of Industrial Minerals.* Greenville, South Carolina, 1988. 25–37.

Brooks, C.K. 1985. Vertical crustal movements in the Tertiary of central East Greenland: a continental margin at a hot-spot. *Z. Geomorph.* Supp. Vol. 54, 101–17.

Brown, E.H. 1960. *The relief and drainage of Wales.* Cardiff.

Budel, J. 1982. (trans. L. Fisher and D. Busche). *Climatic Geomorphology.* Princeton University Press, Princeton, NJ.

Bull, W.B. and Cooper, A.F. 1986. Uplifted marine terraces along the Alpine Fault, New Zealand. *Science,* 234, 1125–8.

Bullard, F.M. 1977. *Volcanoes of the Earth.* 2nd edn. Austin, University of Texas Press.

Burke, K. and Wells, G.L. 1989. Trans-African drainage system of the Sahara: Was it the Nile? *Geology,* 17, 743–7.

Butler, R.F., Gehrels, G.E., McClelland, W.C., May, S.R. and Klepacki, D. 1989. Discordant palaeomagnetic poles from the Canadian Coast Plutonic Complex: regional tilt rather than large-scale displacement? *Geology,* 17, 691–4.

Butt, C.M. 1985. Granite weathering and silcrete formation on the Yilgarn Block, Western Australia. *Aust. J. Earth Sci.,* 32, 415–33.

Campbell, I.B. and Claridge, G.G.C. 1988. Landscape evolution in Antarctica. *Earth Sci. Rev.* 25, 345–53.

Carey, S.W. 1958. The tectonic approach to continental drift. *Continental Drift: A Symposium.* Geology Department, University of Tasmania, Hobart.

Carey, S.W. 1976. *The Expanding Earth.* Elsevier, Amsterdam.

Carey, S.W. 1988. *Theories of the Earth and Universe.* Stanford University Press, California.

Cassie, R.A. 1978. Palaeomagnetic studies of the Suva Marl. BSc (Hons), University of Sydney.

Cerveny, P.F., Naeser, N.D., Zeitler, P.K., Naeser, C.W. and Johnson, N.M. 1988. History of uplift and relief of the Himalaya during the past 18 million years. Evidence from fission-track ages of detrital zircons from sandstones of the Siwalik Group, in Kleinspehn, K.L. and Paola, C. (eds), *New Perspectives in Basin Analysis.* Springer-Verlag, New York. 43–61.

Chappell, J. 1974. Geology of coral terraces, Huon Peninsula, New Guinea: a study of Quaternary tectonic movements and sea level changes. *Bull. Geol. Soc. Amer.,* 85, 555–70.

Chester, D.K. and Duncan, A.M. 1982. The interaction of volcanic activity in Quaternary times upon the evolution of the Alcantara and Simeto Rivers, Mount Etna, Sicily. *Catena,* 9, 319–42.

Chenet, P.Y., Colletta, B., Letouzey, J., Desforges, G., Ousset, E. and Zaghloul, E.A. 1987. Structures associated with extensional tectonics in the Suez rift, in Coward et al. 1987, 551–58.

Chorley, R.J. 1962. Geomorphology and general systems theory. *U.S. Geol, Surv. Prof. Pap.,* 500-B.

Chorley, R.J. 1965. A re-evaluation of the geomorphic system of W.M. Davis, in Chorley, R.J. and Haggett, P. (eds), *Frontiers of Geographical Teaching.* Methuen, London. 21–38.

Chorley, R.C. 1978. Bases for theory in geomorphology, in Embleton, C., Brunsden, D. and Jones, D.K.C. (eds) *Geomorphology. Present problems and future prospects.* Oxford University Press, Oxford. 1–13.

Chubb, L.J. 1957. The pattern of some Pacific Island chains. *Geol. Mag.,* 94, 221–8; and 1961, 98, 170–1, and 99, 278–83.

Clague, D.A., Dalrymple, G.B. and Moberly, R. 1975. Petrology and K–Ar ages of dredged volcanic rocks from the western Hawaii Ridge and the southern Emperor Seamount Chain. *Bull. Geol. Soc. Am.,* 86, 991–8.

Coaldrake, J.E. 1962. The coastal sand dunes of southern Queensland. *Proc. Roy. Soc. Queensland,* 72, 101–6.

Coe, K. 1975. The Hurry Inlet granite and related rocks of Liverpool Land, east Greenland. *Gronl. Geol. Unders.,* 115-, 5–33.

Condie, K.C. 1989. *Plate Tectonics and Crustal Evolution.* 3rd edn. Pergamon, Oxford.

Coney, P.J., Jones, D.L. and Monger, J.W.H. 1980. Cordilleran suspect terranes. *Nature,* 228, 329.

Cook, P.J., Colwell, J.B., Firman, J.B., Lindsay, J.M., Schwebel, D.A. and Von Der Borch, C.C. 1977. Late Cainozoic sequence of South East of South Australia and Pleistocene seal level changes. *BMR J. Geol. Geophys.,* 2, 81–8.

Copeland, P. and Hanson, T.M. 1990. Episodic rapid uplift of the Himalayas revealed by $^{40}Ar/^{39}Ar$ analysis of detrital K-feldspar and muscovite, *Bengal Fan. Geology,* 18, 354–7.

Cornacchia, M. and Dars, R. 1983. Un trait structural majeur du continent africain. Les linéaments centrafricaines du Cameroun au golfe d'Aden. *Bull. Geol. Soc. Fr.,* 25, 101–9.

Cotton, C.A. 1944. *Volcanoes as Landscape Forms.* Whitcombe and Tombs, Christchurch.

Coude-Gaussen, G. 1981. Les Serras da Peneda et do Geres. *Memorias do Centro de Estudos Geograficos,* No. 5. University of Lisbon.

Courtillot, V., Armijo, R. and Tapponnier, P. 1987. Kinematics of the Sinai triple junction and a two-phase model of Arabia–Africa rifting, in Coward et al. (1987) q.v. 559–73.

Coward, M.P., Dewey, J.F. and Hancock, P.L. (eds) 1987. *Continental Extensional Tectonics.* Geological Society Special Publication 28.

Cox, A. and Hart, R.B. 1986. *Plate tectonics: how it works.* Blackwell, Oxford.

Cox, K.G. 1989. The role of mantle plumes in the development of continental drainage patterns. *Nature,* 342, 873–6.

Craig, D.H. 1988. Caves and other features of Permian karst in San Andres Dolomite, Yates Field Reservoir, West Texas, in James, N.P. and Choquette, P.W. (eds) 1988, q.v. Ch. 16, 342–63.

Craig, M.A. 1984. The Permian geology and physiography and landscape evolution of northeastern Victoria. Unpubl. MSc thesis, University of New England, Armidale, Australia.

Crickmay, C.H. 1933. The later stages of the cycle of erosion. *Geol. Mag.,* 70, 337–47.

Crickmay, C.H. 1975. The hypothesis of unequal activity, in Melhorn W.N. and Flemal, R.C. (eds) *Theories of Landform Development,* George Allen and Unwin, London. 103–10.

Dalziel, I.W.D. 1986. Collision and Cordilleran orogenesis: an Andean perspective. pp. 389–404 in Coward, M.P. and Ross, A.C. (eds) 1986. *Collision Tectonics.* Geological Society Special Publication 19.

Dardis, G.F. and Moon, B.P. (eds) 1988. *Geomorphological Studies in Southern Africa.* Balkema, Rotterdam.

Davies, P.J. 1975. Shallow seismic structure of the continental shelf, southeast Australia. *J. Geol. Soc. Aust.* 22, 345–59.

Davis, G.H. 1984. *Structural Geology of Rocks and Regions*. Wiley, New York.

Davis, W.M. 1899. The geographical cycle. *Geogr. J.*, 14, 481–504.

De Jong, K.A. and Scholten, R. (eds) 1973. *Gravity and Tectonics*. Wiley, New York.

De Sitter, L.U. 1964. *Structural Geology*. 2nd edn. McGraw-Hill, New York.

Desrochers, A. and James, N.P. 1988. Early Palaeozoic surface and subsurface paleokarst: Middle Ordovician Carbonates, Mingan Islands, Quebec, in James, N.P. and Choquette, P.W. (eds) 1988, q.v. Ch. 9, 183–210.

De Voto, R.H. 1988. Late Mississippian Paleokarst and related mineral deposits, Leadville Formation, Central Colorado. in James, N.P. and Choquette, P.W. 1988, q.v. Ch.13, 278–305.

De Wit, M.C.J. 1988. Aspects of the geomorphology of the North-Western Cape, South Africa, in Dardis, G.F. and Moon, B.P. (eds) 1988 q.v. 57–69.

Dixon, J.C. and Young R.W. 1981. Character and origin of deep arenaceous weathering mantles on the Bega batholith, southeastern Australia. *Catena*, 8, 97–109.

Dohrenwend, J.C., Wells, S.G., McFadden, L.D. and Turrin, B.D. 1987. Pediment dome evolution in the eastern Mohave Desert, California, in Gardiner, V. (ed.) *International Geomorphology 1986 Part II*. Wiley, Chichester, 1047–51.

Dumitru, T.A., Hill, K.C., Coyle, D.A., Duddy, I.R., Foster, D.A., Gleadow, A.J.W., Green, P.F., Kohn, B.P., Laslett, G.M. and O'Sullivan, A.J. In press. Fission track thermochronology: application to continental rifting of south-eastern Australia. *APEA Journal*.

Dunkelman, T.J., Karson, J.A. and Rosendahl, B.R. 1988. Structural style of the Turkana Rift, Kenya. *Geology*, 16, 258–61.

Durrance, E.M., Bromley, A.V., Bristow, C.M., Heath, M.J. and Penman, J.M. 1982. Hydrothermal circulation and post-magmatic changes in granites of south-west England. *Proc. Ussher Society*, 5, 304–20.

Eardley, A.J. 1963. Relation of uplifts to thrusts in Rocky Mountains. *Am. Assoc. Petrol. Geolog.* Mem 2, 209–19.

Eaton. G.P. 1987. Topography and origin of the southern Rocky Mountains and Alvarado Ridge, in Coward et al. (eds) 1987, q.v. 355–69.

Ebinger, C.J. 1989. Tectonic development of the western branch of the East African rift system. *Bull. Geol. Soc. Amer.*, 101, 885–903.

Elter, P. and Trevisian, L. 1973. Olistostromes in the tectonic evolution of the Northern Apennines, in De Jong, K.A. and Scholten, R. (eds) *Gravity and Tectonics*. Wiley, New York.

Engeln, G.B. 1963. Gravity tectonics in the northwestern Dolomites (N. Italy). *Geologica Ultraaiectina*, no. 13. Rijksuniversiteit te Utrecht.

Evanoff, E. 1990. Early Oligocene paleovalleys in southern and central Wyoming: evidence of high local relief on the late Eocene unconformity. *Geology*, 18, 443–6.

Evamy, D.D., Haremboure, J., Kamerling, P., Knaap, W.A., Moly, F.A. and Rowlands, P.H. 1979. Hydrocarbon habitat of Tertiary Niger Delta. *Bull. Am. Assoc. Petrol. Geol.*, 62, 1–39.

Exley, C.S. 1958. Magmatic differentiation and alteration in the St Austell granite. *Q.J. Geol Soc. Lond.*, 114, 196–227.

Eyles, C.H. and Eyles, N. 1989. The upper Cenozoic White River 'Tillites' of southern Alaska: subaerial slope and fan delta deposits in a strike-slip setting. *Bull. Geol. Soc. Amer.* 101, 1091–102.

Fairbridge, R.W. and Finki, C.W. 1978. Geomorphic analysis of the rifted cratonic margins of Western Australia. *Z. Geomorph.*, 22, 369–89.

Falvey, D.A. and Mutter, J.C. 1981. Regional plate tectonics and the evolution of Australia's passive continental margins. *BMR J. Aust. Geol. Geophys.*, 6, 1–29.

Farhoudi, G. In press. The morphogenetic phase of the Alborz Mountain Ranges of northern Iran. *Z. Geomorph.*

Finkl, C.W. 1979. Stripped (etched) landsurfaces in southern Western Australia. *Aust. Geogr. Studies*, 17, 35–5.

Finkl, C.W. and Fairbridge, R.W. 1979. Paleogeographic evolution of a rifted cratonic margin: S.W. Australia. *Palaeogeog. Palaeoclim. Palaeoecol.* 26, 221–52.

Fischer, K. 1989. The landforms of the German Alps and the Alpine Foreland. *Catena*, Supp. 15, 69–83.

Fisher, W.L. and McGowen, J.H. 1967. Depositional systems of the Wilcox Group of Texas and their relationship to occurrences of oil and gas. *Gulf Coast Assoc. Geol. Socs.* Texas Trans.17, 105–25.

Flenley, J.R. 1984. Andean guide to Pliocene–Quaternary climate. *Nature*, 311, 702–3.

Ford, D.C. 1989. Paleokarst in Canada. in Bosak et al. 1989, q.v. 313–16.

Ford, T.D. 1984. Paleokarst of Britain. in Bosak et al. 1989, q.v. 51–70.

Francis, P. 1976. *Volcanoes*. Penguin, Harmondsworth.

Frankl, E.J. and Cordry, E.A. 1967. The Niger delta province: recent developments onshore and offshore. *World Petrol. Cong.*, 7th, Mexico, Proc. 1B, 195–209.

Friedman, G.M. 1987. Vertical movements of the crust: Case histories from the northern Appalachian Basin. *Geology*, 15, 1130–3.

Galloway, R.W. 1987. The age of landforms in southeastern Queensland. CSIRO Institute of Biological Resources, Division of Water and Land Resources Technical Memorandum 87/2.

Gansser, A. 1983. The morphogenic phase of mountain building, in Hsu, K.J. (ed.) *Mountain building processes*. Academic Press, London. 221–8.

Garrett, S.W. and Storey, B.C. 1987. Lithospheric extension on the Antarctic Peninsula during Cenozoic subduction, in Coward et al. 1987, q.v.419–31.

Gilbert, G.K. 1877. Report on the Geology of the Henry Mountains, in *US Geographical and Geological Survey of the Rocky Mountains Region* (Powell).

Gilchrist, A.R. and Summerfield, M.A. 1990. Differential denudation and flexural isostasy in formation of rifted-margin upwarps. *Nature*, 346, 739–42.

Glazek, J. 1989. Paleokarst of Poland, in Bosak et al. 1989, q.v. 77–105.

Gleadow, A.J.W. and Duddy, I.R. 1987. Fission track analysis of crystalline basement rocks from the Namaqualand mobile belt, South Africa. *Geotrack Report* No. 68, 1–4.

Godard, A. 1982. Les bourrelets marginaux des hautes latitudes. *Bull. Assoc. Geogr. Franc.* 1982. 489, 239–69.

Grabert, H. 1971. Die Prae-Andine Drainage des Amazonas Strompsystems. *Muenster Forsch. Geol. Palaeontol.* 20, 51–60.

Graf, W.L. (ed.) 1987 Geomorphic systems of North America. Boulder, Colorado. *Geol. Soc. America*, Centennial Special Volume 2.

Graf, W.L., Hereford, R., Laity, J. and Young, R.A. 1987. Colorado Plateau. in Graf, W.L. (ed.) 1987, q.v.

Grove, J. 1988. *The Little Ice Age*. Routledge, London.

Guilcher, A. 1988. *Coral Reef Geomorphology*. Wiley, New York.

Gustavson, T.C. and Budnik, R.T. 1985. Structural influences in geomorphic processes and physiographic features, Texas Panhandle: technical issues in siting a nuclear-waste repository. *Geology*, 13, 173–6.

Hack, 1965. Geomorphology of the Shenandoah Vaslley, Virginia and West Virginia and origin of medieval ore deposits. *US Geol. Surv. Prof Paper* 484.

Hack, J.T. 1960. Interpretation of erosional topography in humid temperate regions. *Am. J. Sci.* 258, 80–97.

Hackman, B.D., Charsley, T.J. Key, R.M. and Wilkinson, A.F. 1990. The development of the East African Rift system in north-central Kenya. *Tectonophysics.* 184, 189–211.

Haggerty, J.A. Schlanger, S.O. and Silva, I.P. 1982. Late Cretaceous and Eocene volcanism in the southern Line Islands and implications for hotspot theory. *Geology*, 10, 433–7.

Hall, A. 1986. Deep weathering patterns in north-east Scotland and their geomorphological significance. *Z. Geomorph.* 30, 407–22.

Hall, A. 1987. Weathering and relief development in Buchan, Scotland, in Gardiner, V. (ed.) *International Geomorphology 1986*, 991–1005.

Hall, J. 1859. *Natural History of New York. Palaeontology, 3.* Vol.3, Appleton, New York.

Hamilton, W. 1987. Crustal extension in the Basin and Range Province, southwestern United States. in Coward, Dewey and Hancock (eds) q.v. 155–6.

Hamilton, W. and Myers, B. 1967. The nature of batholyths. *U.S. Geol. Surv.* Prof. Paper 554C, 1–30.

Haq, B.U., Hardenbol, J., and Vail, P.R. 1987. Chronology of fluctuating sea levels since the Triassic. *Science*, 235, 1156–67.

Haworth, R.J. and Ollier, C.D. 1992, in press. The Evolution of the Clarence River, New South Wales. *Earth Surface Processes and Landforms.*

Hawthorne, J.B. 1975. Model of a kimberlite pipe, in Ahrens, L.H., Dawson, J.B., Duncan, A.R. and Erklank, A.J. (eds) *Physics and Chem. of the Earth.* Pergamon, Oxford.

Hays, J. 1967. Land Surfaces and laterites in the north of the Northern Territory, in Jennings, J.N. and Mabbutt, J.A. (eds) *Landform Studies from Australia and New Guinea.* ANU Press, Canberra. Ch.9, pp. 182–210.

He, C. 1987. The characteristics of karst geomorphology in Guizhou, in Gardiner, V. *International Geomorphology 1986. 1095–108.*

Heck, F.R. 1989. Mesozoic extension in the southern Appalachians. *Geology*, 17, 711–14.

Hepworth, J.V. 1964. Explanation of the geology of Sheets, 19, 20, 28 and 29 (Southern West Nile). *Geol. Surv. Uganda*, Report No. 10.

Higgins, C.G. 1975. Theories of landscape development: a perspective, in Melhorn, W.N. and Flemal, R.C. 1975 q.v. 1–28.

Hills, E.S. 1975. *Physiography of Victoria.* Whitcombe and Tombs, Melbourne.

Holm, D.K. and Wernicke, B. 1990. Black Mountains crustal section, Death Valley extended terrain, California. *Geology*, 18, 520–23.

Holmes, A. 1965. *Physical Geology.* Nelson, London.

Hooghiemstra, H. 1984. *Vegetational and climatic history of the High Plains of Bogota, Colombia: a continuous record of the last 3.5 million years.* Cramer, Liechtenstein.

Howell, D.G. 1989. *Tectonics of suspect terranes: mountain building and continental growth.* Chapman and Hall, London.

Hsu, K.J. 1972. When the Mediterranean dried up. *Sci. Am.*, December, 27–36.

Hsu, K.J. (ed.) 1982. *Mountain Building Processes.* Academic Press, London.

Huber, N.K. 1990. The Late Cenozoic evolution of the Tuolumno River, central Sierra Nevada. *Bull. Geol. Soc. Am.*, 102, 102–15.

Idnurm, M. and Senior, B.R. 1978. Palaeomagnetic ages of Late Cretaceous and Tertiary weathered profiles in the Eromanga Basin, Queenslands. *Palaeogeog. Palaeoclimatol. Palaeoecol.*, 24, 263–77

Iijima, A. 1972. Latest Cretaceous–Early Tertiary Lateritic Profile in northern Kitakami Massif, northeast Honshu, Japan. *J. Fac. Sci.* University of Tokyo, 18 325–70.

Illies, J.H. 1972. The Rhine Graben rift system — Plate tectonics and transform faulting. *Geophys., Surveys*, 1, 27–60.

Isaac, K.P. 1983. Tertiary lateritic weathering in Devon, England, and the Palaeogene continental environment in South West England. *Proc. Geol. Assoc.*, 94, 105–114.

Jacobs, E.O. and Thwaites, R.N. 1988. Erosion surfaces in the southern Cape, South

Africa, in Dardis, G.F. and Moon, B.P. (eds) *Geomorphological Studies in Southern Africa.* q.v. 47–55.

James, N.P. and Choquette, P.W. 1988 *Paleokarst.* Springer-Verlag, New York.

Jenkins, D.A.L. 1974. Detachment tectonics in western Papua New Guinea. *Bull, Geol. Soc. Am.*, 85, 533–48.

Jenks, W.F. 1975. Peru, in Fairbridge, R.W. (ed.) 1975. *The Encyclopedia of World Regional Geology, Part 1: Western Hemisphere.* 426–33.

Jennings, J.N. 1971 *Karst.* ANU Press, Canberra.

Jenny, H. 1941. *Factors of Soil Formation. A system of quantative pedology.* McGraw Hill, New York.

Jones, O.T. 1931. Some episodes in the geological history of the Bristol Channel region. *Rept. Brit. Ass. Adv. Sci.*, 57–82.

Judson, S. 1975. Evolution of Appalachian topography, in Melhorn, W.N. and Flemal, R.C. (eds) 1975, q.v. 29–44.

Jutson, 1934. The physiography (geomorphology) of Western Australia. 2nd edn. *Geol. Surv. W.A. Bull.* 95.

Kahle, C.F. 1988. Surface and subsurface paleokarst, Silurian Lockport, and Peebles Dolomites, Western Ohio, in James, N.P. and Choquette, P.W. (eds) 1988, q.v. Ch.11, 229–55.

Karner, G.D. and Weissel, J.K. 1984. Thermally induced uplift and lithospheric flexural readjustment of the eastern Australian highlands. *Geol. Soc. Aust. Abstracts* 12, 293–294.

Kaye, C.A. 1967. Kaolinisation of bedrock of the Boston, Massachusetts area. *U.S. Geol Surv. Prof. Pap.* 575-C:C165–C172.

Kear, D. 1957. Erosional stages of volcanic cones as indicators of age. *New Zealand Journal of Science and Technology*, B38, 671–82.

Kerrans, C. and Donaldson, J.A. 1988. Proterozoic palaeokarst profile, Dismal Lakes Group, N.W.T., Canada, in James, N.P. and Choquette, P.W. 1988, q.v. Ch.8, 167–82.

King, B.C., Le Bas, M.J. and Sutherland, D.S. 1972. The history of the alkaline volcanoes and extrusive complexes of eastern Uganda and western Kenya. *J. Geol. Soc. Lond.*, 128, 173–205.

King, H. 1989. *The Rocks Speak.* Australian Institute of Mining and Metallurgy, Monograph No. 15. Melbourne.

King, L.C. 1955. Pediplanation and Isostacy: an example from South Africa. *Q. J. Geol. Soc. Lond.*, 111, 353–9.

King, L.C. 1967. *The Morphology of the Earth.* Oliver and Boyd, Edinburgh.

King, L.C. 1983. *Wandering Continents and Spreading Sea Floors on an Expanding Earth.* Wiley, Chichester.

Kleeman, J.D. 1984. The anatomy of a tin mineralising A-type granite, in Flood, P.G. and Runnegar, B. (eds) *New England Geology.* Geology Dept, University of New England. 327–34.

Korsch, R.J. 1982. Mount Duval: geomorphology of a near-surface granite diapir. *Z. Geomorph.* 26, 151–62.

Kroonenberg, S.B., Bakker, J.G.M. and van der Wiel, A.M. 1990. Late Cenozoic uplift and paleogeography of the Colombian Andes: constraints on the development of high Andean biota. *Geol. en Mijn.* 69, 279–90.

Kroonenberg, S.B. and Melitz, J.P. 1983. Summit level, bedrock control and the etchplain concept in the basement of Surinam. *Geol. en Mijn.* 62, 389–99.

Kukal, Z. 1984. *Atlantis in the light of modern research.* Elsevier, Amsterdam.

Ladd, H.S. and Hoffmeister, J.E. 1945. *Geology of Lau, Fiji.* Bernice P. Bishop Museum, Bull. 181.

Lambeck, K. and Stephenson, R. 1988. The post-Palaeozoic uplift history of south-eastern *Australia. Aust. J. Earth Sci.*, 33, 253–270.

Langford, F.F. 1974. A supergene origin for vein-type uranium ores in the light of the Western Australian calcrete-carnotite deposits. *Economic Geology.* 69, 516–26.

Le Coeur, C. 1988. Late Tertiary warping and erosion in Western Scotland. *Geogr. Ann.* 70(A), 361–7.

Le Coeur, C. 1989. La question des alterites profondes dans la région des Hébrides internes (Ecosse occidentale). *Z. Geomorph.* Supp. Vol. 72, 109–24.

Le Masurier, W.E. and Rex, D.C. 1983. Rates of uplift and the scale of ice level instabilities recorded by volcanic rocks in Marie Byrd Land, West Antarctica, in Oliver, R.L., James, P.R. and Jago, J.B. (eds) *Antarctic Earth Science,* 663–70. *Aust. Acad. Sci.* Canberra.

Lehman, T.M. 1990. Paleosols and the Cretaceous/Tertiary transition in the Big Bend region of Texas. *Geology, 18,* 362–4.

Lensen, G. 1968. Analysis of progressive fault displacement during downcutting at the Branch River Terrace, South Island, New Zealand. *Bull. Geol. Soc. Am.* 79, 545–56.

Li, Y., Geissman, J.W., Nur, A., Ron, H., and Huang, Q. 1990. Palaeomagnetic evidence for counterclockwise block rotation in the north Nevada rift region. *Geology,* 18, 79–82.

Lidmar-Bergstrom, K. 1982. Pre-Quaternary geomorphological evolution in southern Fennoscandia. *Sveriges Geologiska Undersokning,* 75, 6.

Lidmar-Bergstrom, K. 1988. Exhumed Cretaceous landforms in south Sweden. *Z. Geomorph.,* Supp. Vol. 72, 21–40.

Lillegraven, J.A. and Ostresh, L.M. 1990. Late Cretaceous (earliest Campanian/Maastrichian) evolution of westerly shorelines of the North American Western Interior Seaway in relation to known mammalian faunas. *Geol. Soc. Am.* Spec. Pap. 243, 1–30.

Liu, S. and Zhong, X. 1987. The geomorphology of the Hengduan Mountains, China, in Gardiner, V. (ed.) *International Geomorphology 1986,* Part II. Wiley, Chichester. 229–37.

Liu, T. and Yuan, B. 1987. Megageomorphic features and history of the Chinese loess in Gardiner, V. (ed.) *International Geomorphology 1986,* Part II. Wiley, Chichester. 241–53.

Liu, Z. 1987. Development and filling of caves in Dragon Bone Hill at Zhoukoudian, Beijing, in Gardiner, V. (ed.) *International Geomorphology 1986,* Part II. Wiley, Chichester. 1125–41.

Locardi, E. 1985. Neogene and Quaternary Mediterranean Volcanism: the Tyrrhenian Example, in Stanley, D.J. and Wezel, F.C. (eds) *Geological evolution of the Mediterranean Basin.* Springer-Verlag, New York, Ch. 13 273–91.

Locardi, E. 1988. The origin of the Apenninic arcs. *Tectonophysics,* 146, 105–23.

Lowry, D.C. 1970. The geology of the Western Australian part of the Eucla Basin. *Western Austr. Geol. Surv. Bull.* 122.

Lowry, D.C. and Jennings, J.N. 1974. The Nullarbor Karst, Australia *Z. Geomorph.* 18, 35–81.

Lynn, H.B., Hale, L.D. and Thompson, G.A. 1981. Seismic reflections from the basal contacts of Batholiths. *J. Geophys. Res.,* 86, 10633–8.

Mabbutt, J.A. 1988. Land-surface evolution at the continental time-scale: an example from interior Western Australia. *Earth-Science Revs,* 25, 457–66.

Macumber, P.G. 1977. Permian glacial deposits, tectonism, and the evolution of the Loddon Valley. *Mining and Geology Journal* [Victoria]. 7, 2–4.

Madole, R.F., Bradley, W.C., Loewenherz, D.S., Ritter, D.F., Rutter, N.W. and Thorn, C.E. 1987. Rocky Mountains, in Graf, W.L. (ed.) q.v.

Magill, J. and Cox, A. 1981. Post-Oligocene tectonic rotation of the Oregon Western Cascade Range and the Klamath Mountains. *Geology,* 9, 127–31.

Maignien, R. 1966. *Review of research on laterites.* UNESCO, Paris.

Mann, P. and Burke, K. 1984. Cenozoic rift formation in the northern Caribbean. *Geology*, 12, 732–6.

Mann, P. and Corrigan, J. 1990. Model for late Neogene deformation in Panama. *Geology*, 18, 558–62.

Marbut, C.F. 1928. A scheme for soil classification. *Proc. 1st. Int. Cong. Soil Sci.*, 4, 1–51.

Martin, H. 1975. Structural and palaeogeographical evidence for an Upper Paleozoic sea between southern Africa and South America, in Campbell, K.S.W. (ed.) *Gondwana Geology*. ANU Press, Canberra. 37–59.

Martini, J.E.J. 1981. Early Proterozoic paleokarst of the Transvaal, South Africa. *Proc. 8th Int. Cong. Speleol.*, Part 1, Huntsville, Alabama, 4–5.

Maund, J.G., Rex, D.C., Le Roex, A.P. and Reid, D.L. 1988. Volcanism on Gough Island: a revised stratigraphy. *Geol.Mag.*, 125, 175–181.

Mazzanti, R. and Trevisan, L. 1978. Evoluzione della rete idrografica nell'appennino centro-settentrionale. *Geogr. Fis. Dinam. Quat.*, 1, 55–62.

McCartan, L., Tiffney, B.H., Wolfe, J.A., Ager, T.A., Wing, S.L., Sirkin, L.A.Q., Ward, L.W. and Brooks J. 1990. Late Tertiary floral assemblage from upland gravel deposits of the southern Maryland coastal plain. *Geology*, 18, 311–14.

McMillan, N.J. 1973. Shelves of Labrador Sea and Baffin Bay, Canada. in McCrossan, R.G. (ed.), *The future petroleum provinces of Canada — their Geology and Potential.* Canadian Soc. Petrol. Geol. Mem.1, pp. 473–517.

Melhorn, W.N. and Edgar, D.E. 1975. The case for episodic continental-scale erosion surfaces: a tentative geodynamic model, in Melhorn, W.N. and Flemal, D.E. 1975. *Theories of Landform Development.* George Allen and Unwin, London, 243–76.

Melhorn, W.N. and Flemal, D.E. 1975. *Theories of Landform Development.* George Allen and Unwin, London.

Meyerhoff, H.A. and Olmsted, E.W. 1963. The origins of Appalachian drainage. *Am. J. Sci.*, 332, 21–41.

Meyers, W.J. 1988. Paleokarstic features in Mississippian limestone, New Mexico, in James, N.P. and Choquette, P.W. 1988, q.v. Ch. 14, 306–28.

Millar, C.E., Turk, L.M. and Forth, H.D. 1966. *Fundamentals of Soil Science.* Wiley, New York.

Miller, R.P. 1937. Drainage Lines in bas-relief. *J. Geol.*, 45, 432–8.

Milnes, A.G. 1987. Tectonic evolution of the southern Andes, Tierra del Fuego: a summary, in Schaer and Rodgers (eds) 1987. Ch.9, 173–7.

Milnes, A.R. Bourman, R.P. and Northcote, K.H. 1985. Field relationships of ferricretes and weathered zones in southern south Australia: a contribution to 'laterite' studies in Australia. *Aust. J. Soil Res.*, 23, 441–65.

Molina, E., Cantano, M., Vicente, M.A. and Garcia Rodriguez, P. 1990. Some aspects of palaeoweathering in the Iberian Hercynian Massif. *Catena*, 17, 333–46.

Molnar, P. and Stock, J. 1987. Relative motion of hotspots in the Pacific, Atlantic and Indian Oceans since late Cretaceous time. *Nature*, 327, 587–91.

Moore, D.G., Curray, J.R., Raitt, R.W., and Emmel, F.J. 1974. Stratigraphic-seismic section correlations and implications to Bengal Fan history, in von der Borch, C.C., Sclater, J.G. et al., *Init. Rept. Deep Sea Drilling Project*, 22. US Govt Printing Office, Washington DC 403–12.

Mortimer, C. 1973. The Cenozoic history of the southern Atacama Desert Chile, *Jl. Geol. Soc. Lond.*, 129, 505–26.

Mortimer, N. and Coleman, R.G. 1985. A neogene structural dome in the Klamath Mountains, California and Oregon. *Geology*, 13, 253–6.

Muhs, D.R., Thorson, R.M., Clague, J.J., Mathews, W.H., McDowell, P.F. and Kelsey, H.M. 1987. Pacific coast and mountains system. in Graf, W.L. (ed.) q.v. 1987.

Mussman, W.J., Montanez, I.P. and Read, J.F. 1988. Ordovician Knox paleokarst unconformity, Appalachians, in James and Choquette, 1988, q.v. Ch.10, 210–29.

Myers, J.S. 1975. Vertical crustal movement of the Andes in Peru. *Nature*, 254, 672–4.

Nikiforoff, C. 1949. Weathering and soil evolution. *Soil Sci.*, 67, 2129–30.

Nott, J. 1990. The fluvial landscape history of the middle Shoalhaven basin of southeast New South Wales from 45 Ma to the present, in Gale, S.J. (ed.) 1990. *Long-term Landscape Evolution in Australia (Abstracts)*. Department of Geography and Planning, University of New England, Armidale. 12–13.

Oberlander, T. 1965. *The Zagros Streams: a new interpretation of transverse streams in an orogenic zone.* Syracuse Geographical Studies No. 1. Department of Geography, Syracuse, New York.

Ollier, C.D. 1981. *Tectonics and Landforms.* Longman, London.

Ollier, C.D. 1982a. The Great Escarpment of eastern Australia: tectonic and geomorphic significance. *J. Geol. Soc. Aust.*, 29, 13–23.

Ollier, C.D. 1982b. Geomorphology and tectonics of the Dorrigo Plateau, N.S.W. *J. Geol. Soc. Aust.*, 29, 431–5.

Ollier, C.D. 1983. Weathering or hydrothermal alteration? *Catena*, 10, 57–9.

Ollier, C.D. 1984. Geomorphology of the South Atlantic Volcanic Islands. Part 1: The Tristan da Cunha Group. *Z. Geomorph.*, 28, 367–82.

Ollier, C.D. 1985. Lava Flows of Mount Rouse, Western Victoria. *Proc. R. Soc. Vict.*, 97, 167–74.

Ollier, C.D. 1988a. *Volcanoes.* Blackwell, Oxford.

Olier, C.D. 1988b. A National Park Survey in Western Samoa: Terrain classification on tropical volcanoes, in Applied Geomorphological Mapping: methodology by example. *Z. Geomorph.* Supp. 40, 103–24.

Ollier, C.D. 1992. Tectonics and landscape evolution in southeast Asutralia. *Earth Surface Processes.* In press.

Ollier, C.D., Chan, R.A., Craig, M.A., and Gibson, D.L. 1988a. Aspects of landscape history and regolith in the Kalgoorlie region. *BMR J. Geol. Geophys.*, 10, 309–321.

Ollier, C.D. and Galloway, R.W. 1990. The laterite profile, ferricrete and unconformity. *Catena*, 17, 97–109.

Ollier, C.D., Gaunt, G.F.M. and Jurkowski, I. 1988. The Kimberley Plateau, Western Australia. A Precambrian erosion surface. *Z. Geomorph.*, 32, 239–46.

Ollier, C.D. and Joyce, E.B. 1987. *Regolith Terrain Units of the Hamilton 1:1 000 000 Sheet Area*, Western Victoria. Bureau of Mineral Resources, Geology and Geophysics, Record 1986/33, p.55 + map.

Ollier, C.D. and Marker, M.E. 1980. The Great Escarpment of southern Africa. *Z. Geomorph., Supp. Bd.* 54, 37–56.

Ollier, C.D. and Pain, C.F. 1980. Actively rising surficial gneiss domes in Papua New Guinea. *J. Geol. Soc. Aust.*, 27, 33–44.

Ollier, C.D. and Pain, C.F. 1981. Active gneiss domes in Papua New Guinea: new tectonic landforms. *Z. Geomorph.*, 25, 133–45.

Ollier, C.D. and Pain, C.F. 1988. Morphotectonics of Papua New Guinea. *Z. Geomorph.* Supp. Vol. 69, 1–16.

Ollier, C.D. and Powar, B. 1985. The Western Ghats and the Morphotectonics of Penninsular India. *Z. Geomorph. Supp. Bd.*, 54, 57–70.

Ollier, C.D. and Rajaguru, S.N. 1989. Laterite of Kerala, India. *Geografia Fisica e Dinamica Quaternaria*, 12, 27–33.

Ollier, C.D. and Stevens, N.C. 1989. The Great Escarpment in Queensland. in Le Matire, R.W. (ed). *Pathways in Geology — Essays in Honour of Edwin Sherbon Hills.* Distributor: Blackwells Scientific, Melbourne.

Ollier, C.D. and Taylor, D. 1988. Major geomorphic features of the Kosciusko–Bega area. *BMR J. Geol. Geophys.*, 10, 357–62.

Osborne, R.A.L. 1987. How old are the caves?, in Galloway, R.W. *The age of* landforms in eastern Australia. CSIRO Institute of Biological Resources, Division of Water and Land Resources. Technical Memorandum 87/2. 53–4.

Osborne, R.A.L. 1986. Cave and landscape chronology at Timor Caves, New South Wales. *J. and Proc. R. Soc. New South Wales*, 119, 55–75.

Osborn, F.A.L. and Branagan, D.F. 1988. Karst landscapes of New South Wales, Australia. *Earth-Science Revs.* 25, 467–80.

Osmaston, M.F. 1973. Limited lithosphere separation as a main cause of continental basins, continental growth and epeirogeny, in Tarling, D.N. and Runcorn, S.K. (eds) *Implications of Continental Drift to the Earth Sciences*. Academic Press, London. 649–74.

Ota, Y. 1985. Marine terraces and active faults in Japan with special reference to co-seismic events, in Morisawa, M. and Hack, J.T. (eds) *Tectonic Geomorphology*. Allen and Unwin, Boston. Ch. 15, pp. 345–66.

Padilla, L.E. 1981. Geomorphología de posibles areas peneplanizadas en la Cordillera Occidentale de Colombia. *Mem. lo Sem. Cuat. Colombia*, Revista CIAF 6, 391–402.

Pain, C.F. 1985. Morphotectonics of the continental margins of Australia. *Z. Geomorph. Supp. Bd.* 54, 23–36.

Pain, C.F. and Ollier, C.D. 1984. Drainage patterns and tectonics around Milne Bay, Papua New Guinea. *Rev. Geom. Dyn.*, 32, 113–20.

Palmquist, R.C. 1975. The compatibility of structure lithology and geomorphic models, in Melhorn, W.N. and Flemal, R.C. (eds) 1975, q.v., 145–68.

Parrish, R.R. 1983. Cenozoic thermal evolution and tectonics of the Coast Mountains of British Columbia: 1 Fission-track dating, apparent uplift rates, and pattern of uplift. *Tectonics*, 2, 601–31.

Partridge, T.C. and Maud, R.R. 1987. Geomorphic evolution of southern Africa since the Mesozoic. *South African J. Geol.* 90, 179–208.

Partridge, T.C. and Maud, R.R. 1988. The geomorphic evolution of southern Africa: a comparative review, in Dardis, G.F. and Moon, D.P. (eds) 1988. q.v. 5–15.

Peirce, H.W. Damon, P.E. and Shafiqullah, M. 1979. An Oligocene (?) Colorado Plateau edge in Arizona. *Tectonophysics*, 61, 1–24.

Penck, A. and Bruckner, E. 1909. *Die Alpen im Eiszeitalter*. Tauchnitz, Leipzig.

Peulvast, J.P. 1988. Pre-glacial landform evolution in two coastal high latitude mountains: Lofoten–Vesterålen (Norway) and Scoresby Sund area (Greenland). *Geografiska Annaler*, 70A, 351–60.

Pfeffer, K.H. 1989. The karst landforms of the northern Franconian Jura between the Rivers Pegnitz and Vils. *Catena*, Supp. 15, 253–60.

Pickering, S.M. and Hurst, V.J. 1989. Commercial kaolins in Georgia. in Fritz, W.J. (ed.) *Excursions in Georgia Geology*. Georgia Geological Society Guidebooks. 9, 29–75.

Pierce, K.L. 1965. Geomorphic significance of a Cretaceous deposit in the Great Valley of Pennsylvania. *US Geol. Sur. Prof. Paper 525*, C152–6.

Pierce, W.G. 1957. Heart Mountain and South Fork detachment thrusts of Wyoming. *Am. Assoc. Petrol. Geologists*, Bull, 41, 591–626.

Pieri, M. and Groppi, G. 1981. *Subsurface geological structure of the Po Plain, Italy*. C.N.R. Pub. 414. Progetto Finalizzato Geodinamica. Milan.

Pigram, C.J. and Davies, H.L. 1987. Terranes and the accretion history of the New Guinea orogen. *BMR J. Aust. Geol. Geophys.* 10, 193–211.

Pillans, B. 1990. Pleistocene marine terraces in New Zealand: a review *NZ J. Geol. Geophys.*, 33, 219–31.

Pillans, B. and Walker, P. In press. Basalt and bulldust: soils and landscape evolution near Nimmitabel, southern Monaro, NSW, in Williams, De Decker and Kershaw (eds) in press. q.v.

Potter, P.E. 1978. Significance and origin of big rivers. *J. Geol.*, 86, 13–33.

Pugh, J.C. 1955. Isostatic readjustment and the theory of pediplanation. *Q. J. Geol. Soc. Lond.* 111, 361–9.

Purdy, E.G. 1974. Reef configurations: cause and effect, in Laporte, L.F. (ed.) *Reefs in Space and Time.* Soc. Econ. Palaeontologists and Mineralogists, Spec. Pub. 18, 9–76.

Quennell, A.M. 1959. Tectonics of the Dead Sea. *Int. Geol. Cong. 20 (Mexico),* 385–405.

Radwanski, S.A. and Ollier, C.D. 1959. A study of an East African catena. *J. Soil Sci.,* 10, 149–68.

Rai, R.K. 1987. Evidences of rejuvenation of the Deccan foreland, India, with particular reference to the Meghalaya Plateau, in Gardiner, V. (ed.) *International Geomorphology 1986,* Part II. Wiley, Chichester. 255–63.

Raman, P.K. 1981. Geomorphic evolution and its significance in exploration planning of the East Coast Bauxite deposits of India. *J. Geol. Soc. India.* 22, 488–95.

Reeves, F. 1946. Origin and mechanics of thrust faults adjacent to the Bearpaw Mountains, Montana. *Bull. Geol. Soc. Am.* 57, 1033–47.

Rodgers, J. 1983. The life history of a mountain range — the Appalachians, in Hsu, K.J. (ed.) 1982. *Mountain Building Processes.* Academic Press, London. 229–41.

Rowlands, N.J., Blight, P.G., Jarvis, D.M. and von der Borch, C.C. 1980. Sabka and playa environments in late Proterozoic grabens, Willonoran Ranges, South Australia. *J. Geol. Soc. Aust.,* 27, 55–68.

Royden, L.H. and Burchfiel, B.C. 1987. Thin-skinned N–S extension within the convergent Himalayan region: gravitational collapse of a Miocene topographic front, in Coward et al. (eds) 1987, q.v. 611–19.

Rutten, M.G. 1969. *The Geology of Western Europe.* Elsevier, Amsterdam.

Said, R. 1981. *The Geological Evolution of the River Nile.* Springer-Verlag, New York.

Sando, W.J. 1988. Madison Limestone (Mississippian) paleokarst: a geological geologic synthesis, in James, N.P. and Choquette, P.W. 1988, q.v. Ch.12, 256–77.

Sangster, D.F. 1979. Plate tectonics and mineral deposits: a view from two perspectives. *Geoscience Canada,* 6, 185–8.

Schaer, J.P. 1987. Evolution and structure of the High Atlas of Morocco, in Schaer and Rodgers (eds) 1987. Ch.7, 107–27.

Schaer, J.P. and Rodgers, J. 1987. *The Anatomy of Mountain Ranges.* Princeton University Press, Princeton, NJ.

Schmidt, K.H. 1989. Geomorphology of limestone areas in the northeastern Rhenish Slate Mountains. *Catena,* Supp. 15, 165–77.

Schmidt, P.W. Currey, D.T. and Ollier, C.D. 1976. Sub-basaltic weathering, damsites, palaeomagnetism and the age of lateritisation. *J. Geol. Soc. Aust.,* 23, 267–70.

Schmidt, P.W. and Ollier, C.D. 1988. Palaeomagnetic dating of Late Cretaceous to Early Tertiary weathering in New England, NSW, Australia, *Earth-Science Reviews,* 25, 363–72.

Schmidt, P.W., Prasad, V. and Raman, P.K. 1983. Magnetic ages of some Indian laterites. *Palaeogeog. Palaeoclim. Palaeoecol.* 44, 185–202.

Schoch, R.M. 1989. *Stratigraphy Principles and Methods.* Van Nostrand Reinhold, New York.

Scholl, D.W. and Marlow, M.S. 1974. Global tectonics and the sediments of modern trenches: some different opinions, in Kahle, C.F. (ed.) 1974. Plate Tectonics — Assessments and Reassessments. *Am. Assoc. Petrol. Geol. Memoir* 23, 255–72.

Scholl, D.W., Marlow, M.S. and Cooper, A.K. 1977. Sediment subduction and offscraping at Pacific margins, in M. Talwani and W.C. Pitman (eds), 1977. *Island Arcs, Deep Sea Trenches and Back-Arc Basins.* American Geophysical Union, Washington, 199—210.

Scholz, C.H. 1977. Transform fault systems of California and New Zealand: similarities in their tectonic and seismic styles. *J. Geol. Soc. Lond.*, 133, 215–29.

Schumm, S.A. and Lichty, R.W. 1965. Time, space and causality in geomorphology. *Am. J. Sci.* 263, 110–19.

Shea, J.H. 1982. Twelve fallacies of uniformitarianism. *Geology*, 10, 455–60.

Sheppard, S.M.F. 1977. The Cornubian batholith, SW England: D/H and $^{18}O/^{16}O$ studies of kaoline and other alteration minerals. *Q.J. Geol. Soc. Lond.* 133, 573–92.

Shilts, W.W., Aylsworth, J.M., Kaszycki, J.M. and Klassen, R.A. 1987. Canadian Shield, in Graf, W.L. (ed.) 1987 *Geomorphic systems of North America: Boulder, Colorado.* Geol. Soc. America, Centennial Special Volume 2, 119–61.

Smith, A.G. 1982. Late Cenozoic uplift of stable continents in a reference frame fixed to South America. *Nature*, 296, 400–4.

Smith, B.J. and McAlister, J.J. 1987. Tertiary weathering environments and products in northeast Ireland, in Gardiner, V. (ed.) *International Geomorphology 1986*, Part II. Wiley, Chichester. 1007–31.

Sparks, B.W. 1972. *Geomorphology.* Longman, London.

Stanley, R.G. 1987. New Estimates of displacement along the San Andreas fault in central California based on paleobathymetry. *Geology.* 15.

Starkel, L. 1987. The Role of inherited forms in the present-day relief of the Polish Carpathians, in V. Gardiner (ed.) *Int. Geom. 86*, Part II. Wiley, Chichester. 1033–45.

Stewart, A.J. Blake, D.H. and Ollier, C.D. 1986. Cambrian river terraces and ridgetops in Central Australia: oldest persisting landforms? *Science*, 23, 758–61.

Stewart, J.H. 1983. Extensional tectonics in the Death Valley area, California: Transport of the Panamin Range structural block 80 km northwestward. *Geology*, 11, 153–7.

Stirling, M.W. 1990. The Old Man Range and Garvie Mountains: tectonic geomorphology of the Central Otago peneplain, New Zealand. *NZ J. Geol. Geophys.*, 33, 233–343.

Stoddart, D.R. 1969, *Climatic geomorphology: review and reassessment.* Progress in Geography 1. Edward Arnold, London.

Strahler, A.N. 1950. Equilibrium theory of erosional slopes approached by frequency distribution analysis. *Am. J. Sci.*, 248, 673–96, 800–14.

Sullivan, W. 1974. *Continents in Motion.* Macmillan, London.

Summerfield, M.A. and Thomas, M.F. 1987. Long-term landform development: editorial introduction, in Gardiner, V. (ed.) *International Geomorphology 1986*, Part II. Wiley, Chichester. 927–33.

Suppe, J. 1987. The active Taiwan mountain belt, in Schaer and Rodgers (eds) 1987. Ch. 15, 277–93.

Sutherland, F.L. 1977. Zeolite minerals in the Jurassic dolerites of Tasmania: their use as possible indicators of burial depth. *J. Geol. Soc. Aust.* 24, 171–8.

Sutherland, F.L. and Wellman, P. 1986. K–Ar ages of Tertiary volcanic rocks, Tasmania. *Papers and Proc. R. Soc. Tasmania*, 120, 77–86.

Swann, D.H. 1963. Classification of Genevievian and Chesterian (Late Mississippian) rocks of Illinois. *Ill. Geol. Surv. Rept.* 216.

Szatmari, P. 1983. Amazon rift and Pisco–Jurua fault: their relation to the separation of North America from Gondwana. *Geology*, 11, 300–4.

Tabbutt, K.D. 1990. Temporal constraints on the tectonic evolution of Sierra de Famatina, northwestern Argentina, using the fission track method to date tuffs interbedded in synorogenic clastic sedimentary strata. *J. Geol.*, 98, 557–66.

Tardy, Y. and Nahon, D. 1985. Geochemistry of laterites, stability of Al-goethite, A-hematite and Fe (III)-Kaolinite in bauxites and ferricretes: an approach to the mechanism of concretion formation. *Am. J. Sci.*, 285–903.

Thomas, M.J. 1989. The role of etch processes in landform development II: etching and the formation of relief. *Z. Geomorph.*, 33, 257–274.

Thomas, M.F. and Summerfield, M.A. 1987. Long-term landscape development: key themes and research problems, in Gardiner, V. (ed.) *International Geomorphology* 1986, Part II. Wiley, Chichester. 935–56.

Thornbury, W.D. 1969. *Principles of Geomorphology.* 2nd edn. Wiley, New York.

Tingey, R.J. 1985. Uplift in Antarctica. *Z. Geomorph.*, Supp. Vol. 54, 85–99.

Trumpy, R. 1980. *An Outline of the Geology of Switzerland.* Wepf, Basel.

Trumpy, R. 1983. Alpine Paleogeography: a reappraisal. in Hsu, K.J. (ed.) 1982. Mountain Building Processes, Academic Press, London. 149–56.

Twidale, C.R. 1990. The origin and implications of some erosional landforms. *J. Geol.*, 98, 343–64.

Twidale, C.R. and Harris, W.K. 1977. The age of Ayers Rock and the Olgas, central Australia, *Trans. R. Soc. S. Aust.*, 101, 45–50.

Ufimtsev, G.F. 1990. Morphotectonics of the Mongolia–Siberian mountain belt. *J. Geodynamics*, 11, 309–25.

Urrutia-Fucugauchi, J. 1981. Paleomagnetic evidence for tectonic rotation of northern Mexico and the continuity of the Cordilleran orogenic belt between Nevada and Chihuahua. *Geology*, 9, 178–83.

Van de Graaf, W.J.E. 1981. Palaeogeographic evolution of a rifted cratonic margin — discussion. *Palaeogeog. Palaeoclim, Palaeoecol.* 34, 163–72.

Vera, J.A. Ruiz-Ortiz, P.A., Garcia-Hernandez, M. and Molina, J.M. 1988. Paleokarst and related pelagic sediments in the Jurassic of the Subbetic Zone, Southern Spain, in James, N.P. and Choquette, P.W. (eds) 1988. q.v. Ch.17, 354–84.

Wager, L.R. 1937. The Arun river drainage pattern and the rise of the Himalaya. *Geogr. J.*, 89, 239–49.

Wallace, C.A., Lidke, D.J. and Schmidt, R.G. 1990. Faults of the central part of the Lear and Clark line and fragmentation of the late Cretaceous foreland basin in west-central Montana. *Bull. Geol. Soc. Am.*, 102, 1021–37.

Webb, J.A. 1989. The age and development of the Chillagoe karst, north Queensland. *Australian and New Zealand Geomorphology Group, Fourth Conference.* Abstracts 34–5.

Webb, J.A. Finlayson, B.L., Fabel, F.G. and Ellaway, M. In press. The geomorphology of the Buchan area — implications for the history of the south eastern highlands of Australia, in Williams, De Decker and Kershaw (eds) in press, q.v.

Wellman, P. 1979. On the isostatic compensation of Australian topography. *BMR J. Aust. Geol. Geophys.*, 4, 373–82.

Wellman, P. 1986. Intrusions beneath large intraplate volcanoes. *Exploration Geophys.*, 17, 135–9.

Wellman, P. 1987. Eastern highlands of Australia: their uplift and erosion. *BMR J. Aust. Geol. Geophys.* 10, 277–86.

Wellman, P. 1988. Tectonic and denudational uplift of Australian and Antarctic highlands. *Z. Geomorphologie*, 32, 17–29.

Wellman, P. and McDougall, I. 1974. Cainozoic igneous activity in eastern Australia. *Tectonophysics*, 23, 49–65.

Wernicke, B. 1981. Low-angle normal faults in the Basin and Range Province: Nappe tectonics in an extending orogen. *Nature*, 291, 645–8.

Wezel, F.C. 1982a. The structure of the Calabro-Sicilian Arc: result of a post-orogenic intra-plate deformation, in Leggett, J.K. (ed.) *Trench-Forearc Geology: Sedimentation and Tectonics on Modern and Ancient Active Plate Margins.* Geol. Soc. London. Spec. Publ. No. 10, 1982. 345–54.

Wezel, F.C. 1982b. The Tyrrhenian Sea: a rifted krikogenic-swell basin. *Mem. Soc. Geol. It.*, 24, 531–68.

Wezel, F.C. 1988. A young Jura-type fold belt within the Central Indian Ocean. *Boll. Oceanologia teorca ed Applicata*, 6, 75–90.

Williams, G.E. 1989. Tidal rhythmites: geochronometers for the ancient Earth–Moon system. *Episodes*, 12, 162–71.

Williams, G.W. 1983. *The Tertiary auriferous alluvial deposits of north central Victoria*. Bur, Min. Resour. Geol. Geophys. Rec. 1983/27, 137–43.

Williams, M.A.J., De Decker, P. and Kershaw, A.P. (eds) In press. *The Cainozoic in Australia: a re-appraisal of the evidence*. Geol. Soc. Aust. Spec. Publ.

Williams, P.W. (ed.) 1988. The Geomorphology of plate boundaries and active continental margins. *Z. Geomorph., Supp. Bd.* 69.

Woldegabriel, G., Aronson, J.L. and Walter, R.C. 1990. Geology, geochronology and rift basin development in the central sector of the Main Ethiopia Rift. *Bull. Geol. Soc. Am.*, 102, 439–58.

Woodward, L.A. 1977. Rate of crustal extension across the Rio Grande Rift. *Geology*, 5, 269–72.

Woolnough, W.G. 1927. Pres. Address, Pt 1: The chemical criteria of peneplanation. Pt 2: The duricrust of Australia. *J. Proc. R. Soc. NSW.* 61, 17–53.

Wright, R.L. 1963. Deep weathering and erosion surfaces in the Daly River basin, Northern Territory. *J. Geol. Soc. Aust.*, 10, 151–64.

Wright, V.P. 1988. Paleokarst and paleosols as indicators of paleoclimate and porosity evolution: a case study from the Carboniferous of South Wales, in James, N.P. and Choquette, P.W. 1988. q.v.

Wu, F.T. and Wang, P. 1988. Tectonics of western Yunnan Province, China. *Geology*, 16, 153–7.

Wynne Edwards, H.R. 1977. Metallogenic implications of the millipede model — ductile ensiallic orogenesis in the Proterozoic (abstr.) Orthodoxy and Creativity at the Frontiers of Earth Science. University of Tasmania, Hobart.

Yim, W.W.S. In press. Tin placer genesis in northeastern Tasmania, in Williams, De Decker and Kershaw (eds) In press, q.v.

Yuan, D. 1987. New observations on tower karst. in Gardiner, V. (ed.) *International Geomorphology 1986*, Part II. 1109–23.

Zeil, W. 1979. The Andes: a geological review. *Beitr. Reg. Geol. Erde*, Borntraeger, Berlin, Stuttgart. 13, 260.

Zorin, Y.A. 1981. The Baikal Rift: an example of the intrusion of asthenospheric material into the lithosphere as the cause of disruption of lithospheric plates. *Tectonophysics*, 73, 91–104.

Index

absolute dating 13
active margins 122, 123, 125, 209
Alaska 24, 198
Alps (European) 195, 197
Amazon 41
Andes 41, 94, 111, 157, 158, 185, 187
Antarctica 99, 105, 106, 188
antecedent drainage 29
Apennines 197
Appalachians 14, 33, 69, 85, 95, 124, 184, 195
Arizona 158
Arun River 31, 32
Athens 5
Atlantic 105, 179
Atlas Mountains 168
atolls 132
Australia 2, 14, 17, 26, 27, 56, 57, 78, 79, 98, 99, 100, 116, 118, 127, 151, 179, 194
Avon River 33
Ayers Rock 86, 87

ball clay 53
Barron River 39
basalt 4, 21, 72
Basin and Range Province 169, 191
bauxite 60
Bay of Biscay 108
beach ridges 18
Bendigo 20, 21
Bengal Fan 23, 192
Borneo 45
Boston 54
bourrelets marginaux 116, 193
Brahmaputra 23
Brazil 37, 41, 162
breakup unconformities 27, 124
British Columba 84, 113, 170, 185
Bubnoff unit 15
Buchan 52
Buchan Caves 71
buried palaeokarst 69

Calcrete 59
California 115, 169, 176, 198
Canada 19, 29, 30, 69, 185
Cape York 33, 101
carbon dating 13
Caribbean 175
Carpathians 204
Cascade Range 115
caves 68
cave sediments 25
Chile 94, 126, 187
Chillagoe Caves 71
China 30, 72, 73, 113, 192
Clarence River 27, 41, 42
climatic geomorphology 204
coasta ridges 2
coasts 124
Colombia 94
Colorado 69, 83
Columbia River 138, 139
Congo River 46, 47
continental drift 2
continental margins 104, 115
Coober Pedy 11, 76, 91
coral islands 132
core complex 169
Corsica 110
cyclical theories 77, 198
Cyprus 161

Dartmoor 67
Davenport Ranges 21, 22, 85
Dead Sea 107, 173, 176
Death Valley 155, 169
Deccan 138, 139, 162
deep leads 20, 63
deltas 23
dendritic drainage 29
denudation chronology 205
depositional basins 26
depositional landforms 17

Devon 51, 56
dolerite 101
Dolomites 91
domes 162, 198
drainage
  antecedent 29
  barbed 39, 41
  disruption 145
  drowned 45
  patterns 29
  radial 34
  superimposed 33
Drakensberg 119, 193
drowned valleys 45
Dundas dome 36, 37
duricrusts 55, 63, 90
dykes 8, 160
dynamic equilibrium 50, 202

Ebor Volcano 34, 143, 147, 148
economic geology 206
Ecuador 187
Egypt 70
Elburz Mountains 191
emanations 206
Emperor Chain 133, 135
England 20, 35, 56, 66, 67, 81
Epeirogeny 181
erosion surfaces 63, 74
etchplains 64, 88, 205
Ethiopia 173
evolutionary geomorphology 212
exhumed surfaces 96
exotic terranes 41, 110, 111, 112, 198, 209
expanding earth 127

faults 8, 148, 164
  Alpine 176
  Bocono 178
  chasmic 178
  listric 170
  normal 165
  Pisco-Jurua 177
  San Andreas 176
  strike-slip 176
  thrust 165, 170
Fen River 30
ferricrete 51, 56, 58
Fiji 107, 110, 115
Finchhafen Peninsula 26
fission track dating 14, 155, 185, 187, 192
France 56, 81

Ganges 23
Georgia 53, 124, 130
Germany 52, 63, 70, 82
Ghats 193
Gibraltar 130
gipfelflur 95, 96, 97

glacial sediments 23
glaciation 2, 101
Glenelg River 36
gneiss mantled domes 158, 159
Goodenough Island 158, 159
Gondwanaland 93, 106, 116
Gough Island 132, 138
graben 165
gravity tectonics 196
Great Artesian Basin 28
Great Barrier Reef 26
great escarpments 27, 118, 119, 146
granite 8, 154
granodiorite 154
Greenland 19, 99, 188, 195
greybilly 56
Gulf of Arabia 107
Gulf of Mexico 23
Guyana 84, 187

Hamersley 100
Hawaii 133, 135
Himalayas 31, 32, 33, 155, 168, 192, 204
historical geomorphology 205
hornfels 8, 154
horsts 165, 168
hot spots 133, 151
Hunter River 27
hydrothermal alteration 67

ice age 126, 127
Iceland 132, 161
India 59, 60, 94, 168, 193
Indus River 192
intrusions 8, 154
inversion of relief 60, 140, 141, 144
Ireland 26, 61, 68
island arcs 152
Italy 44, 45, 91, 138, 153, 197

Japan 59, 130

Kafu River 40 41,
Kalgoorlie 52, 57, 58, 207
Kalahari 125
Kalimantan 45
Karroo 161, 162
karst 68
Kenya 163, 166
Kimberley 20, 24, 25, 43, 44, 86, 98, 99, 188
Klamath Mountains 198
Kosciusko 78
Kunlun Mountains 192

Lake Albert 40
Lake Baikal 175
Lake District 35, 198
Lake George 165, 167, 169
Lake Kyoga 40, 46
Lake Victoria 40, 46, 47

Lapstone Monocline 86, 87
lateral stream 144
laterite 58
Laurasia 106
Line Islands 135
listric faults 170
little ice age 5
Loddon River 24
loess 18
Lord Howe Island 150
Lough Neagh 26

Madagascar 57, 116, 131
mantle plumes 162
marginal swells 193
marine erosion 149
marine sediments 26
marine terraces 11
Maryland 19
maturity (soil) 50
Mayunmarca 5
McArthur River 207
Meckering Line 119
median plateaux 191
Mediterranean 26, 28, 130, 131
Mendip Hills 20, 25, 81
Messinian Event 28, 46
Mexico 115, 185
Mid-Atlantic Ridge 104
Milne Bay 30, 31, 43
Mississippi 21, 22, 47
Mole Granite 156
Monaro Volcano 20, 34, 146, 147
monoclines 86, 87, 148
Montana 168
moraine 12
Morocco 168
morphotectonics 208
mountain building 2, 182
mountains 180
mountains, classification 182
mountains, border 191
Mount Buffalo 158
Mount Duval 155, 156
Mount Elgon 86, 138
Mount Etna 145
Mount Isa 207
Mount St Helens 5
Muller Range 196
Murray Basin 28, 103
Murrumbidgee 19

necks 159
New Caledonia 131
New Mexico 69, 170
New South Wales 25, 27, 34, 41, 54, 56, 66, 69,
    71, 73, 79, 120, 121, 125, 140, 143, 146,
    147, 148
New York 54

New Zealand 6, 10, 82, 96, 97, 113, 116, 127,
    128, 138, 176, 179
Niger 23
Nile 5, 46, 47, 130, 131
Norway 195
Nullarbor Plain 19, 26, 103

oceans 124,
Ohio 66, 69
Ooldea Range 17
Oregon 115, 198
orogeny 181

Pacific 105, 135
palaeokarst 69
palaeoplains 78
palaeosols 66
palaeomagnetism 13, 18, 51, 88
Pangaea 106
panplains 78
Papua New Guinea 30, 31, 43, 96, 113, 114,
    129, 152, 155, 156, 196
Parana River 41
Passive margins 116
pediplains 78
peneplains 63
Pennsylvania 69, 85
Peru 5, 126, 185, 186
placer deposits 20
planation surfaces 74, 81
planeze 137
plate tectonics 104, 181, 208
plateaux 183
Po Basin 197
Poland 197, 204
Portugal 94
potassium argon dating 4, 13, 53
Pyrenees 107

Qattara Depression 70
Queensland 17, 26, 39, 59, 60, 71

radial drainage 34
Randschwellen 116, 193
Red Sea 174
relict karst 70
resurgent landscapes 24
Rhine 31, 46, 175, 197
rhombochasm 107
Rhone 131
rifts 164
rift valleys 172
Rio Grande 170
river capture 12, 37, 38, 39
river reversal 39
river sediments 19
river terraces 12
Rocky Mountains 84, 171, 184
rotation 115
Ruwenzori Mountains 3, 189, 190

salt karst 73
San Andreas Fault 176
sand dunes 17
sandplains 61, 62
Sardinia 110
Scotland 51, 52, 160
seafloor spreading 105
sea level 15, 126
sediments, cave 25
sediments, glacial 23
sediments, marine 26
sediments, river 19
Sepik River 43
Serro Do Mar 119, 195
Seychelles 131
Shoalhaven River 27
silcrete 56
sills 161
Sierra Nevada 83, 189
Snake River 138
Snowy Mountains 119
soil 13, 49
South Africa 14, 56, 59, 69, 93, 125, 161
South Australia 11, 18, 19, 55, 90
Spain 56, 61, 70
sphenochasms 107
stable isotopes 67
subduction 105, 168, 181
Suez rift 175
Sulawesi 45
Sumatra 45
Sunda Shelf 45, 46
superimposed drainage 33
suspect terranes 111
Sweden 26, 52, 53, 96
systems theory 201

Taiwan 113, 203
Tasmania 20, 101
Tasman Sea 116, 118, 134, 179, 194
Tatra Mountains 197
tectonics geomorphology 208
tephrochronology 6

Texas 69, 70, 73
Thames 46
Tibet 31, 168, 192
till 24
tillite 24
tilt blocks 165, 189
Timor Caves 25, 71, 72
Transantarctic Mountains 119, 188
Troodos Range 161
Tristan da Cunha 131, 149
Tweed Volcano 147, 163, 194
twin lateral streams 141, 144
Tyrrhenian Sea 44, 45, 170, 171

Uganda 2, 3, 40, 41, 64, 65, 75, 86, 139, 163
unequal activity 205
uniformitarianism 210

Vail curves 127, 128
Venezuela 84
Vesuvius 5, 138
Victoria 20, 21, 24, 26, 37, 55, 66, 71, 79, 85, 139, 144
volcanoes 135, 136, 149

Wales 35, 36, 66, 70, 84
Wannon River 148
Warrumbungle Volcano 146, 160, 161
Washington 115, 170
weathering 48, 139, 207
weathering profile 48, 49
weathering sequences 12
weathering, sub-basaltic 54
Western Australia 19, 20, 24, 25, 57, 58, 59, 61, 62, 64, 81, 89, 90, 119
Western Samoa 12
Wyoming 19, 69, 83, 168

Yellowstone 168

Zagros Mountains 191
Zambia 208